Modeling and Simulation in Biomedical Engineering

Modeling and Simulation in Biomedical Engineering

Applications in Cardiorespiratory Physiology

Willem van Meurs

McGraw Hill

New York Chicago San Francisco
Lisbon London Madrid Mexico City
Milan New Delhi San Juan
Seoul Singapore Sydney Toronto

The McGraw·Hill Companies

Cataloging-in-Publication Data is on file with the Library of Congress

Modeling and Simulation in Biomedical Engineering

1 2 3 4 5 6 7 8 9 0 DOC/DOC 1 9 8 7 6 5 4 3 2 1

ISBN 978-0-07-171445-7
MHID 0-07-171445-6

The pages within this book were printed on acid-free paper.

Sponsoring Editor
Michael Penn

Editorial Assistant
Michael Mulcahy

Editorial Supervisor
David E. Fogarty

Production Supervisor
Pamela A. Pelton

Project Manager
Aloysius Raj, Newgen Publishing and Data Services

Copy Editor
Doreen McLaughlin

Proofreader
Ramanathan Deepika

Indexer
Willem van Meurs

Art Director, Cover
Jeff Weeks

Composition
Newgen Publishing and Data Services

À Michou, avec amour

About the Author

Willem van Meurs obtained his MSc degree in electrical engineering at the Eindhoven University of Technology in 1987, and his doctorate in control engineering from Paul Sabatier University, Toulouse, France, in 1991. At the University of Florida, from 1992 to 1998, he was coinventor of the Human Patient Simulator, commercialized by Medical Education Technologies, Inc. (METI), Sarasota, Florida. As a consultant to METI, from 1998 to the present, he was codeveloper of the simulators: PediaSIM®, BabySIM®, iStan®, and METIman™, as well as of the Muse™ software. He was an invited associate professor of applied mathematics at the University of Porto from 1998 to 2001. He was president of the Society in Europe for Simulation Applied to Medicine (SESAM) from 2005 to 2007. Since 1998 he has been an investigator at the *Instituto de Engenharia Biomédica* (INEB) Porto, Portugal, with a program focusing on physiologic models for educational simulation and on a training simulator for obstetric emergencies. He lives with his wife and two children in Lahitte-Toupière, France.

Contents at a Glance

Contents

Foreword

Demand for patient simulation-based education and training in health care continues to accelerate worldwide. Physicians, nurses, pharmacists, therapists, aids, and in fact all health care professionals now learn using various forms of patient simulation as part of their student curricula, post-graduate specialization, and lifelong continuing education and recertification. While inanimate anatomical models have been used for more than half a century, only during the past two decades have computerized mathematical models of human physiology, pharmacology, and pathology brought unparalleled interactive realism to patient simulation.

Our team started working on simulators for training anesthesia residents in 1985 under the mentorship and with active contributions from Dr. Nik (JS) Gravenstein, an anesthesiologist, and Dr. Jan Beneken, an electrical engineer and physiologic modeler. Starting in 1992, Dr. Willem van Meurs transformed our prototype simulator into a fully interactive, model-driven system. We did not know it at the time, but we were applying systems biology and systems pharmacology, and multi-physics modeling, before those terms became popular.

Now, after the commercialization of over 6000 full-body, model-driven acute care simulators, based on or derived from our initial design, there is a clear need for training of a larger international community of engineers who are going to design the next generation of patient simulators. Hence, the critical importance and timely arrival of this text, *Modeling and Simulation in Biomedical Engineering: Applications in Cardiorespiratory Physiology*. This book meets those training needs as well as the needs of biomedical engineers, simulation scholars, industry developers, and research scientists working on other applications involving modeling and simulation of human physiology.

Dr. van Meurs describes the modeling and simulation process in clear steps, breaking down its complexity into logically organized components. He presents a framework for modeling the cardiorespiratory system, and provides detailed examples of conceptual and mathematical models. The final chapter on simulation and simulator design provides an eloquent illustration of how model requirements can be derived from more general process and device requirements. My education and experience as a physician educator with an undergraduate degree in computer science provided me with sufficient background to read, understand, and apply the material presented in this book. I am confident that students and professionals working in the field will find that this text positively impacts their learning and advancement.

MICHAEL L. GOOD, M.D.
Professor of Anesthesiology
Dean, University of Florida College of Medicine
Folks H. Peterson Dean's Distinguished Professor

Preface

Modeling and simulation are now a stand-alone methodology of similar value to laboratory investigations and clinical trials.[1] Yet, a concise introductory text on modeling and simulation and the basic techniques required to implement it in the biomedical field is still missing. This book proposes to fill this void. Part I breaks down what may seem like a rather unwieldy modeling and simulation process into five manageable steps: Requirements, conceptual models, mathematical models, software implementation, and simulation results and validation. In Part II we provide the reader with a framework and a basic set of deterministic, continuous-time models for the cardiorespiratory system. Part III covers the advanced topics: sensitivity analysis and setting model requirements as part of an encompassing simulation and simulator design. For a more detailed overview of the content, we point to the summaries at the beginning of each chapter.

The target audience consists of those studying or working in biomedical engineering: engineers, physicists, applied mathematicians, but also biologists, physiologists, and clinicians. Required background knowledge is not deep, but rather broad, and includes basic knowledge of cardiorespiratory physiology, fluid dynamics, the kinetic theory of gases, differential equations, and some basic programming skills. By the same token, the field and this text act as an integrator of knowledge and skills at a senior undergraduate or (post)graduate level. Depending on the training and experience of the reader, he or she can refer to one of the references for further reading if background knowledge needs to be acquired or refreshed. Concepts from signal processing and control theory on which the text elaborates are introduced.

The first review problem in Chapters 2 to 6 concerns modeling and simulation of the same system, inviting the student to apply the presented theory to a single modeling and simulation project spanning Part I of the book. This project, or a similar one, can be

[1] Hardman JG, Ross JJ: Modelling: a core technique in anaesthesia and critical care research, *Br J Anaesth.* 2006 Nov; 97(5):589–92.

used as a practical assignment accompanying a one-semester theoretical class.

The overall cardiorespiratory block diagram and the models presented in Part II represent an application of the theory presented in Part I. At the same time they represent a simplified version of the cardiorespiratory models used in the Human Patient Simulator (HPS®), co-invented by the author. Clinical educators and simulator technicians using this simulator or derived products are expected to gain further insights in the modeling and simulation process and in specific modeling aspects of these training tools from this book. Their primary focus should be on conceptual models, rather than on mathematical models and software implementation.

Human physiology is exquisitely beautiful and complex, which makes it all the more surprising that we can often capture the essence of its dynamics by using basic physics, mathematics, and engineering techniques. Being able to do so is quite gratifying and potentially life-saving. I hope you will enjoy the ride as much as I do!

WILLEM VAN MEURS

Acknowledgments

McGraw-Hill author and physiologic modeler Lee Waite introduced me to the company. Medical Education Technologies, Inc. CTO Carlos Moreno recommended me and helped me prioritize and organize my work. Taisuke Soda, now at another publisher, and Michael Mulcahy at McGraw-Hill provided encouraging comments and much appreciated assistance.

The Department of Anesthesiology of the University of Florida College of Medicine (1992 to present), Medical Education Technologies, Inc., Sarasota, FL, (1996 to present), and the *Instituto de Engenharia Biomédica*, Porto, Portugal (1998 to present) provide a home to my research and development activities spanning medicine and engineering, and academics and industry. Many clarifications were included in the text after short courses at the Universities of Porto and Zaragoza.

I have had the privilege to personally interact with the following physiologic modelers: Rute Almeida, Hugo Alonso, Peter Andriessen, Tim Antonius, Jim Azukas, Sidarto Bambang Oetomo, Yawen Barowski, João Bernardes, Ron Carovano, Ana Cavaleiro, Nicolas Chauveau, Nicole de Beer, Inês Domingues, Tammy Euliano, Filipe Ferreira, David Gaba, Alexandra Gavina, Jane Goodwin, Marco Grit, Wolfgang Heinrichs, Christine Hélies, Richard Helyer, Pascal Heynen, Devin Johns, Pedro Lago, Pablo Laguna, Samsun Lampotang, Jos Lerou, Mariana Lobo, Charlotte Lommen, Tom Looke, Elisa Magosso, Joaquim Marques de Sá, Mark McClure, Richard Melker, Luis Melo, Teresa Mendonça, Stefan Mönk, Alzira Mota, Bosseau Murray, Patricia Neto, Jeroen Nieuwenhuizen, Erik Nikkelen, Maria Öhrn, Guid Oei, Candida Pinto, Rodolfo Pinto, Mário Rodrigues, Irfan Sajan, Howard Schwid, Ty Smith, José Teixeira, Mark Tooley, Mauro Ursino, Jan van der Aa, Hans van Oostrom, Ton Verbraak, Adriaan Versprille, Denham Ward, Dwayne Westenskow, and Mariken Zijlmans.

Very special thanks to: Jan Beneken, physiologic modeler extraordinaire; Yves Sévely, who introduced me to state variable models; Mike Good, inventor and mentor; Nik (JS) Gravenstein, source of knowledge and inspiration; Lou Oberndorf, visionary entrepreneur; Carla and Pedro Sá Couto, PhD students, competent modelers of the cardiovascular and respiratory systems, respectively; Luísa Bastos, PhD student, searching, researching; Hugo Azevedo, leading full-body model-driven simulator developer; and Diogo Ayres de Campos, exemplary meritocrat.

My yoga teachers Thatcher Morse and Christine Périgaud made me more aware of my own heartbeat and breathing, respectively. My wife, children, family, friends, and fellow musicians keep these rhythms going.

Modeling and Simulation in Biomedical Engineering

CHAPTER 1
Introduction

In this chapter we review or introduce the basic concepts and vocabulary underlying modeling and simulation. Signals are defined and used to distinguish systems from their environment. System properties that play an important role in selecting mathematical formalisms for system modeling, analysis, and simulation are presented. Modeling and simulation are powerful ways of studying systems when direct interaction with the actual system is limited or undesirable. Empirical models can be used to represent input–output behavior. If we also want to represent internal system structure and functioning, an explanatory model is required. Depending on our interest in past or future system behavior, descriptive or predictive models are used. Two examples of modeling and simulation in biomedical engineering are presented. A brief introduction of the symbolic notation used in this book concludes the chapter.

1.1 Signals and Systems

Physical, biological, and social phenomena can be described as events or processes. Events are characterized by occurrence and time of occurrence. Breaking a bone, for example, is an event. Processes involve more gradually evolving entities such as time, length, and temperature. Fracture healing is an example of a process. We can use our own senses to evaluate an entity, for example, temperature as hot or cold, or use a measurement instrument, which, in combination with a unit and a scale, transforms the entity into a quantity with a specific value. For example, the body temperature of this person (quantity) is 36.5 (value) °C (unit). A constant is a quantity that has a specific fixed value. For example, Avogadro's constant is equal to 6.02×10^{23} mol^{-1}. A variable is a quantity that may assume any one of a specified set of values. A continuous variable (CV) can take on an—in principle—infinite number of values in a finite interval, whereas a discrete variable (DV) can only take on a finite number of specific values in a finite interval. These values may be equidistant or not. A special case of a discrete variable is a binary variable, which only takes on the values 0 and 1. A variable can take on values as function of, for example, time or space. A time dependent variable is often called a *signal,* even if no messages are transmitted. The time variable can also be continuous or discrete, leading to continuous-time (CT) and discrete-time (DT) signals, respectively. Figure 1.1 shows several types of signals and common conversions between them. Familiar, but not always rigorously defined designations are included between brackets.

In many phenomena we can distinguish structured ensembles of objects[1] joined in regular interaction or *systems*. A system can be natural or manmade and can be further delineated and described in terms of the signals or variables via which it interacts with the environment. A variable from the environment that influences the evolution of a system is called an *independent* or *input variable*. We sometimes distinguish control inputs and disturbances, depending on whether these independent variables are under our control or not. Disturbances can be measurable or not. Any *dependent* system variable that we are interested in constitutes an *output variable*, whether it influences the environment or not. A system and its input and output variables can be represented in a block diagram (Fig. 1.2).

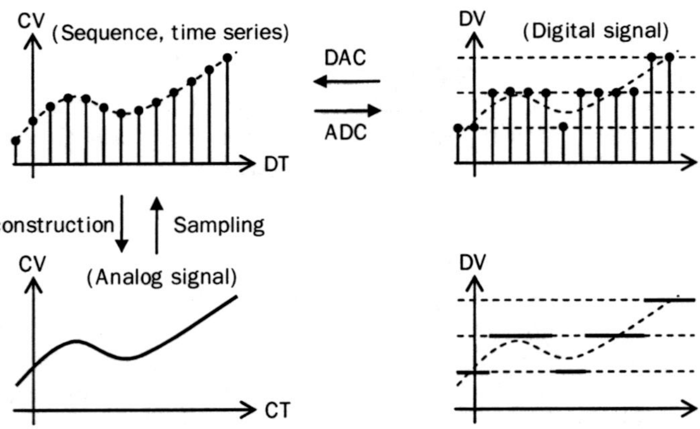

FIGURE 1.1 Signals and common conversions. ADC, analog to digital conversion, involving quantization; DAC, digital to analog conversion. See main text for other abbreviations.

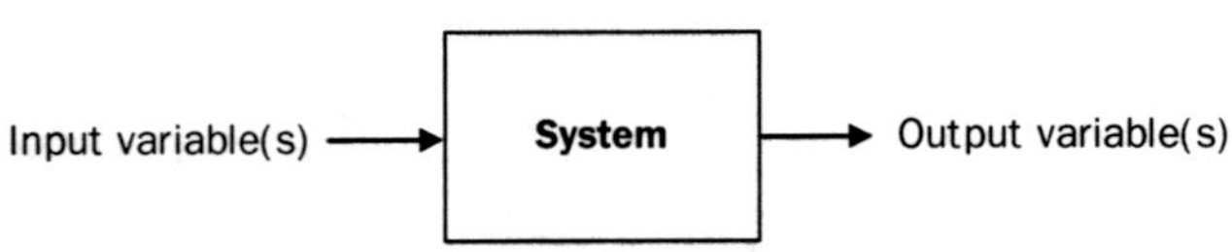

FIGURE 1.2 System block diagram.

[1] We will limit our consideration of autonomous subjects to human-machine interfaces.

For example, if we consider a car as the system, then the positions of the accelerator and the steering wheel are control inputs, and the position of the car is an output variable. But we could also be interested in its speed and acceleration, or the fuel level, which would then be added to the list of output variables. For a more precise description, we would have to take into account the measurable disturbance: wind velocity.

In this section, we defined a variable as a physical quantity that may assume any one of a specified set of values. In subsequent chapters, we will encounter the use of the same term to designate a symbol in a mathematical formula representing a time-varying physical quantity (Chap. 4), a symbol in software code representing a variable or a (constant) parameter in an equation (Chap. 5), and a symbol on the axis of a graph of simulation results (Chap. 6). These are quite different concepts, yet their designation by a single term is at the core of the modeling and simulation process. For example, we can talk about the blood pressure in a graph of simulation results as if it were the original physical quantity, "forgetting" the successive abstractions between the physical phenomenon and the plot.

1.2 System Properties

Once we have defined a system and its boundaries, we can start studying its properties. The first set of system properties are based on input–output variables and play a role in selecting mathematical formalisms for system modeling, analysis, and simulation.

SISO and MIMO

According to the number of input and output variables, we classify systems as single-input, single-output (SISO), multiple-input, multiple-output (MIMO) (Fig. 1.3), or combinations of those: SIMO and MISO.

Continuous-Time and Discrete-Time

We classify systems as continuous-time or discrete-time according to their input and output variables (Fig. 1.1). A system that has a mix of continuous and discrete-time inputs and outputs is called a *hybrid system*. Systems that involve events, rather than processes are often

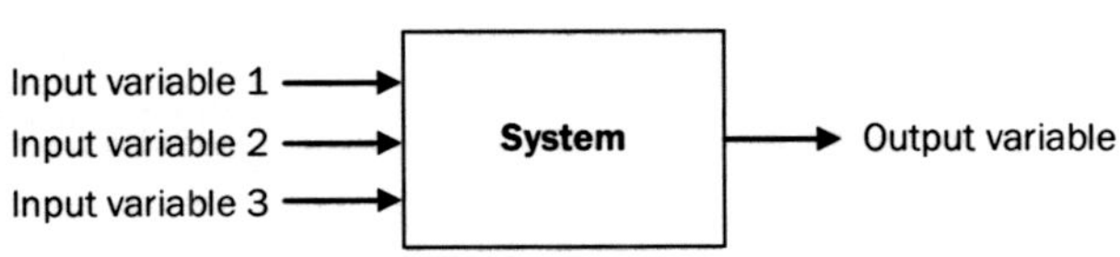

FIGURE 1.3 Block diagram of a MISO system.

called *discrete-event systems* (Law, 2000; Cassandras, 2008). In this book we will focus on modeling and simulation of process type systems.

A second set of system properties is based on system behavior, and once established, plays a primary role in predicting that behavior. These properties also play a role in selecting mathematical formalisms for system modeling, analysis, and simulation. In the definitions below $y(t)$ is the output or response of the system at rest to the input $u(t)$. We will limit ourselves to *causal* systems, for which $y(t_0)$, the output at a specific time t_0, depends only on the past or present input: $u(t)$ with $t \le t_0$, for all $u(t)$ and t_0. This introductory text is also limited to *deterministic* systems for which future outputs are completely determined by the current state of the system and future inputs.[2]

Static and Dynamic

If for all inputs and times t, the output of a system at time t depends on the input at time t only (and not on the input at any other time) we call the system static. If not, we call the system dynamic (Fig. 1.4).

Linear and Nonlinear

Let C be a constant. If the response of the system to the input $C^*u(t)$ is $C^*y(t)$ for all $u(t)$ and C, we call the system homogeneous. Let $y_1(t)$ be the response to input $u_1(t)$ and $y_2(t)$ the response to the input $u_2(t)$. If the response of the system at rest to the input $u_1(t) + u_2(t)$ is $y_1(t) + y_2(t)$ for all $u_1(t)$ and $u_2(t)$, we call the system additive. A system that is both homogeneous and additive is called *linear* (L). If not, the system is nonlinear (NL) (Fig. 1.5).

Time-Invariant and Time-Variant

The input $u(t)$ delayed by a time $t = t_0$ is $u(t - t_0)$. If the response to $u(t - t_0)$ is $y(t - t_0)$ for all $u(t)$ and t_0, the system is time-invariant (TI), if not it is time-variant (TV) (Fig. 1.6).

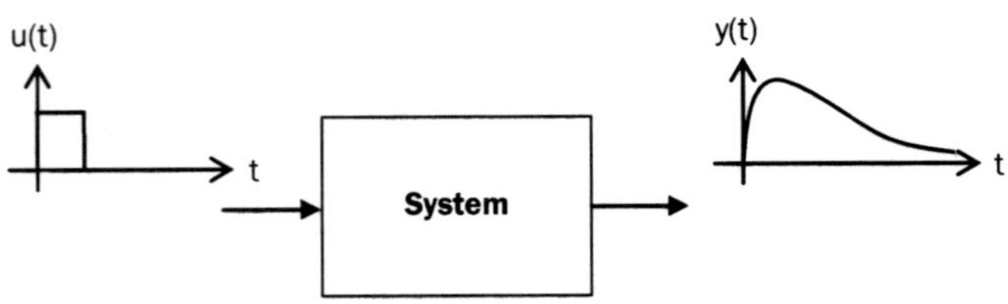

FIGURE 1.4 Response of a dynamic system.

[2] Note that we frequently encounter deterministic systems, such as filters, processing random signals.

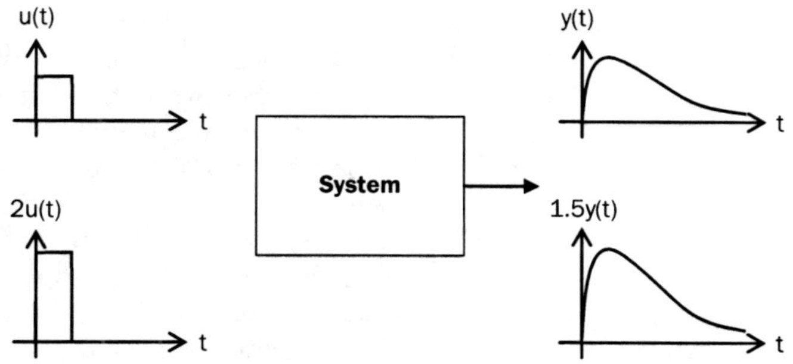

FIGURE 1.5 Responses of a nonlinear system.

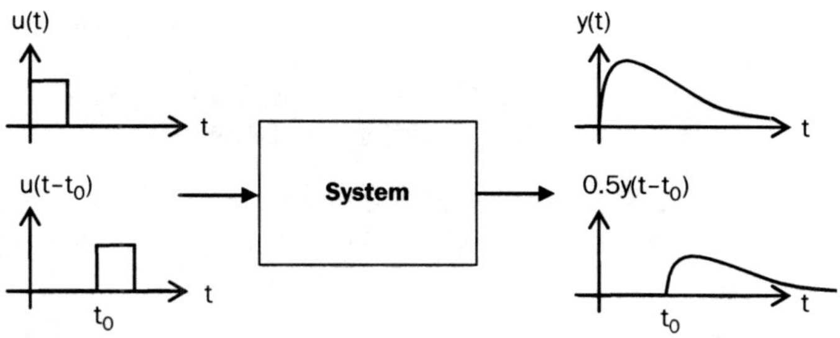

FIGURE 1.6 Responses of a time-variant system.

Systems may be LTI for a subset of inputs, for example, over a certain amplitude range of the input variable(s), or for a limited time. Modeling techniques such as the impulse response and transfer function [see for example Roberts (2003)] are well suited for analysis and design of SISO, LTI systems. But to use these techniques, the input variable needs to be known beforehand, sometimes even in a specific analytical form. This does not make their use very convenient in real-time simulations. Moreover, many physiological systems are MIMO, NL, and TV. In Chap. 4, we will introduce a mathematical description that applies to modeling and simulation of this broader category of systems. This introductory text is limited to *finite dimensional* dynamic systems that can be described by differential equations or difference equations as opposed to systems that are described by partial differential equations.

1.3 Modeling and Simulation

There are various ways to study a system [also see Law (2000)]. "To study" is used here in a broad sense, including gaining a better understanding, system optimization, and training in use or operation. Experimenting with the actual system is acceptable if nobody will get hurt, or if no significant damage is caused if we make a mistake. It is generally not acceptable if the real system is, for example, a nuclear power plant, an airplane, or a patient. In those cases we construct a model.[3] For some systems and experiments, we may be able to construct a physical model, for example, an anatomical model to train needle insertion. For other systems, physical models may be difficult and expensive to construct. Physical models are also difficult to adapt to different situations. In those cases, and if the envisioned application is served by it, we may be able to construct a mathematical model, that is, a mathematical representation of the relationships between system variables that are of interest to us. A "good" mathematical model contains a lot of information about the system, but to use it to study system behavior, we need either an analytical solution, or, if that is not available, a software implementation and a simulation. Before selecting or designing a mathematical model for the response of a dynamic system to an input, two fundamental questions about our interest in the behavior of the system should be asked:

1. Are we merely interested in externally observable behavior, or do we want to understand internal structure and functioning?

The answer to this question determines the type of model at a very basic conceptual level: An interest in internal functioning requires an explanatory, white-box, or knowledge-based model that reflects the structure and functioning of the underlying system (down to a certain depth). An interest limited to external aspects of system behavior may be served effectively and efficiently by an empirical or black-box model, reflecting only essential aspects of response dynamics.

A second fundamental question is:

2. Are we interested in past, present, or future system behavior?

The answer to this question may also influence the type of a model, especially its mathematical formulation, which may reflect a descriptive, real time, or predictive use, even if the underlying conceptual model is the same. Figure 1.7 summarizes these interests and associated model types and uses.

A third, somewhat less fundamental question, but of particular importance in biology and medicine, is if we are interested in the behavior of one or more particular individuals, or in typical behavior for a class or population. The answer to this question may also influence the choice of conceptual or mathematical model, and has an impact on the associated parameter estimation procedure.

[3] In Chapter 6 we will come back to the use of animal models.

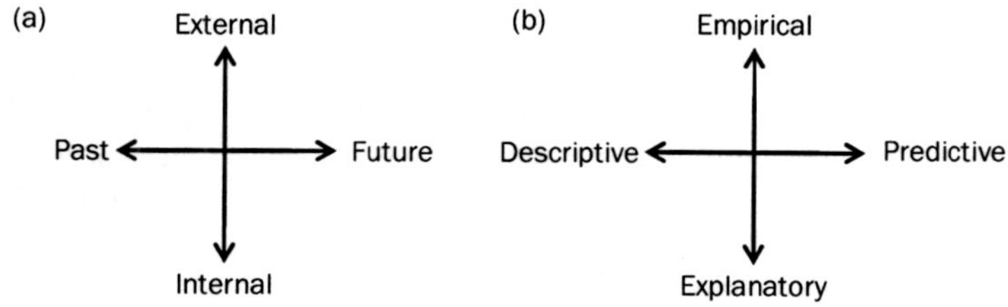

FIGURE 1.7 Interest in system behavior (*a*) and model type and use (*b*).

For example, based on data for the planet Mars gathered by Tycho Brahe (1546–1601), Johannes Kepler (1571–1630) formulated empirical laws that turned out to apply to the movements of all visible planets in the solar system. The first two are:

1. All planets move in elliptical orbits, with the Sun at one focus.
2. A line that connects a planet to the Sun sweeps out equal areas in the plane of the planet's orbit in equal times (Halliday, 2001).

These were first used to describe past movement, and then to predict future planetary motion. Isaac Newton (1643–1727) was able to provide an explanatory model for the same phenomena, based on his law of universal gravitation and his laws of motion. Albert Einstein (1879–1955) provided an alternative explanatory model, based on general relativity and the curvature of space under the influence of the mass of the Sun.

The distinctions in response to the above questions are generally not binary oppositions, but may involve several levels. For example, the deepest level of representation of an explanatory model of an organism may include systems such as the cardiovascular and respiratory systems and corresponding organs, or it may go further and represent tissues, cells, molecules, atoms, etc. Many models used in biomedical engineering are "gray-box" models and combine some level of representation of internal structure and function with empirical, black-box models based exclusively on input–output data. Another example of levels on the axes of Fig. 1.7 is that meteorological and climate models involve similar variables, but descriptions and predictions involve very different time scales. A physical model is also a system, and stretching the meaning of the word "objects" in the definition (Sec. 1.1) a bit, we can even consider a mathematical model a system. Therefore, the system properties introduced in Sec. 1.2 can also be used to further describe models. For example, we can talk about an explanatory deterministic MIMO model of cardiorespiratory dynamics.

1.4 Applications in Biomedical Engineering

Areas in medicine that benefit from modeling and simulation include research, therapy, training, equipment design, and control of devices. Many types of models are used to represent a multitude of physiological systems in very diverse simulation applications. The purpose of this section is to present two examples, rather than to give a systematic review. We will use the terminology introduced in this chapter. Sá Couto et al. (2010) presented a model for use in a training simulator for obstetricians and neonatal intensive care physicians (Fig. 1.8).

The changes in the circulation during and shortly after birth are complex and quite spectacular: Perfusion of the lungs goes from almost nothing to the full cardiac output, the placental circulation is removed, and the ductus arteriosus, a communication between the pulmonary and systemic circulation, necessary for intrauterine life, but potentially fatal to a newborn, closes. Figure 1.9 shows some of the input and output variables of the continuous-time dynamic model representing these hemodynamic phenomena.

The input variables of this MIMO model are set by a discrete event system. A gray-box model explicitly reflects several of the

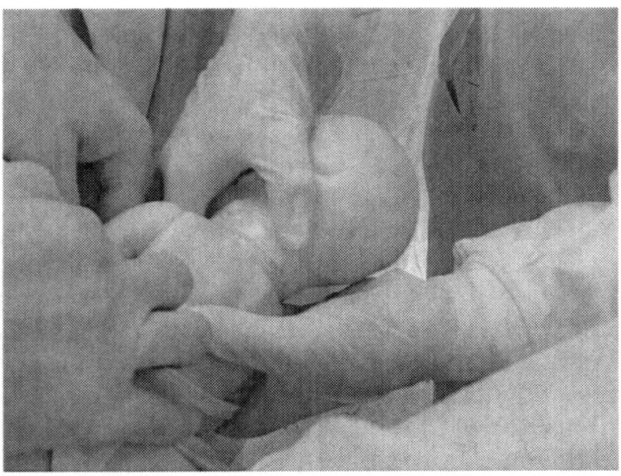

FIGURE 1.8 Prototype delivery simulator. (Picture Dr. Luísa Bastos.)

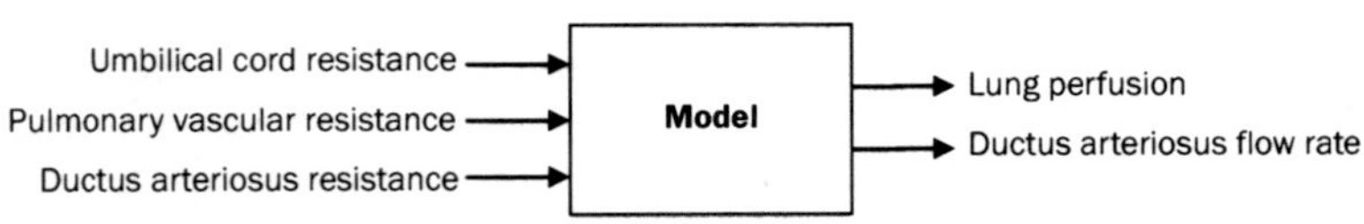

FIGURE 1.9 Block diagram of a model for educational simulation of hemodynamic transitions at birth.

anatomical structures included in the fetal-neonatal circulation, such as the umbilical cord and the ductus arteriosus. The system and mathematical model contain several nonlinearities, for example, those resulting from the cardiac valves. The envisioned model use is real time (neither descriptive nor predictive, Fig. 1.7), and the patient population considered in the article is an average, normal, term fetus.

Intracranial pressure (ICP) is an important variable in the intensive care unit. Increased ICP can result in compression of blood vessels and impede perfusion and oxygenation of the brain. ICP depends on multiple factors, including cerebrospinal fluid dynamics, intracranial compliance, and cerebral hemodynamics. Ursino et al. (1995) presented a mathematical model for the ICP response to a typical clinical test: injection and withdrawal of a few milliliters of saline in the craniospinal space. The primary goal of this continuous-time mathematical model is to understand and explain the complex ICP dynamics. Work carried out using Ursino's publication targets therapeutic and educational applications.

Hardman and Ross (2006) conclude that modeling and simulation are now a stand-alone methodology of similar value to laboratory investigations and clinical trials. The purpose of Part I of this book is to introduce the reader to the modeling and simulation process and to the basic techniques required to implement it, but first we present symbolic notation.

1.5 Symbolic Notation

In a text spanning the clinical world, physiology, physics, chemistry, mathematics, and computer engineering, it is impossible to define a standard symbolic notation for involved quantities that is accepted by all. We propose to optimize notation from an educational, model-building perspective, and retain the following conventions:

- From a mathematical perspective, continuous time is denoted by t, and discrete time by n. We distinguish variables from (constant) parameters by using lower case and capitals, respectively, and by explicitly including time dependency, for example, flow rate $f(t)$, resistance R.
- From the perspective of physics, we use the same symbol for the same quantity. For example, volume, pressure, and flow rate will be designated by $v(t)$, $p(t)$, and $f(t)$, respectively.
- From an anatomo-physiological perspective, we use subscripts to designate a particular location or process. For example, the blood pressure in the aorta is designated by $p_{ao}(t)$. When considering both a gas and a blood phase, we use capital subscripts for gases, and lower case subscripts for

blood, and append a particular gas if necessary. For example, the partial pressures of oxygen in alveolar gas and arterial blood are designated by $p_AO_2(t)$, and $p_aO_2(t)$, respectively.

We recommend keeping symbols meaningful, consistent, and compact. For critical model quantities, such as input and output variables, it is often useful to establish a table including full variable names, symbols, and units. This notation follows several, but not all, of the conventions of a frequently used standard in cardiorespiratory physiology (Nunn, 1993). For example, the primary symbol in our notation represents the physical quantity flow rate without distinguishing between gas and blood.[4] Units and their symbolic notation will be introduced in Chap. 4.

Review Problems

1.1 Define the terms signal, system, model, and simulation.

1.2 Let $y(t)$ be the output of a system at rest to the input $u(t)$. The graphs below show two input–output pairs. Is the system nonlinear or time-variant?

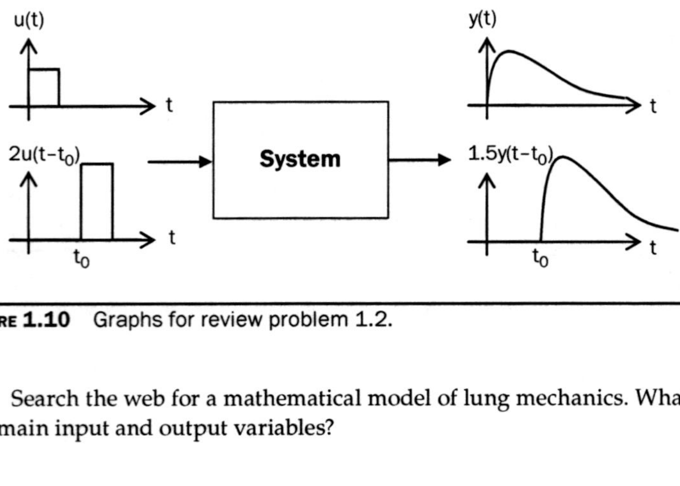

FIGURE 1.10 Graphs for review problem 1.2.

1.3 Search the web for a mathematical model of lung mechanics. What are the main input and output variables?

References and Further Reading

Cassandras CG, Lafortune S. *Introduction to discrete event systems*. 2nd ed. New York: Springer; 2008.

Halliday D, Resnick R, Walker J. *Fundamentals of physics*. 6th ed. New York: John Wiley & Sons; 2001.

Hardman JG, Ross JJ. Modelling: a core technique in anaesthesia and critical care research. *Br J Anaesth*. 2006 Nov; 97(5):589–92.

[4] The mentioned standard represents these quantities by $\dot{V}$ and $\dot{Q}$, respectively.

Law AM, Kelton WD. *Simulation modeling and analysis.* 3rd ed. Boston: McGraw-Hill; 2000.

Nunn JF. *Nunn's applied respiratory physiology.* 4th ed. Oxford: Butterworth Heinemann; 1993.

Roberts MJ. *Signals and systems: Analysis of signals through linear systems.* Boston: McGraw Hill; 2003.

Sá Couto CD, Andriessen P, van Meurs WL, Sá Couto PM, Ayres-de-Campos D. A model for educational simulation of hemodynamic transitions at birth. *Pediatr Res.* 2010; 67(2):158–165.

Ursino M, Iezzi M, Stocchetti N. Intracranial pressure dynamics in patients with acute brain damage: a critical analysis with the aid of a mathematical model. *IEEE Trans Biomed Eng.* 1995 Jun; 42(6):529–40.

PART I

Theory

CHAPTER 2

Model Requirements

In this chapter we introduce a stepwise approach to the modeling and simulation process and present the first step: How to set requirements for a model of the system under study. Qualitative aspects of model design include the input–output variables, conditions to be represented, structures and functions to be included, and the type of model to be used. Quantitative aspects include ranges, resolutions, and bandwidths of input and output variables. Additional requirements address how a model needs to be implemented, and how input and output variables need to be interfaced to the environment. Target response data play a central role in validation of a model before its use to explore certain aspects of system behavior.

The requirements presented in this chapter provide input to all subsequent steps of the modeling and simulation process, and are often further specified in an iterative manner (Fig. 2.1).

In the model development phase, all major steps are executed, up to simulation results and validation. One or more of the steps may be repeated, depending on the results of the validation. Other phases may follow and lead to further iterations, for example, validation by independent experts, comments by model users, expansion of the model, and so on. Model requirements may evolve from one phase to the other.

The novice modeler may want to read this key chapter twice: one cursory initial consideration, followed by a second in-depth reading once the concepts underlying the subsequent steps of the modeling and simulation process are assimilated.

2.1 Qualitative Aspects

Model requirements are often set in dialog with clinical investigators or basic life scientists. At the onset of a modeling and simulation project it is usually clear which organ(s) or global physiological system(s) will be considered. A modeler who does not have a background in these fields may first have to refresh his or her

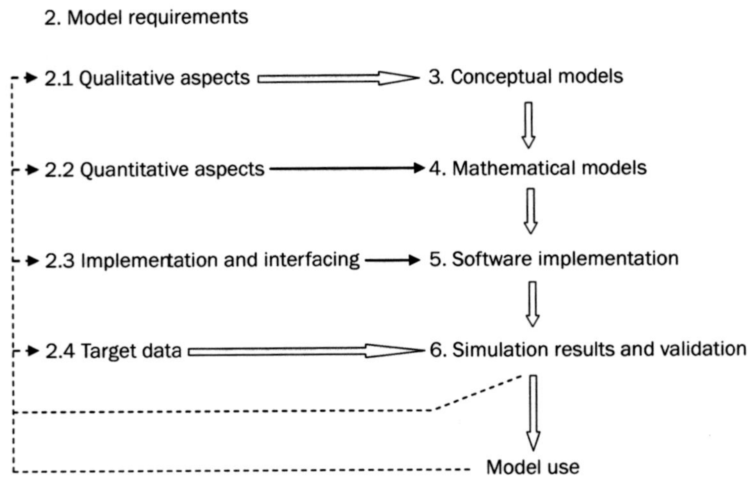

FIGURE 2.1 The modeling and simulation process.

knowledge of relevant physiology (Berne, 2004; Guyton, 2006), and possibly of the health care environment, for example, via a visit to the hospital guided by a clinician. Qualitative choices with potential quantitative implications are those of the population, the physiological state(s) of the subject, and the pathologies to be represented. The population is specified in terms of species, age, gender, height, and weight. A model may be designed for an (hypothetical) average member of the selected population. The "population" may be a specific individual, in which case the model should be designed so that it is possible to estimate model parameters from available data. A special physiological state that almost always requires simulation is the normal resting state of a subject. Body position can be specified as, for example, supine: lying down, face up. Other physiological states are, for example, hypovolemia or hypoxemia. Examples of pathologies are cardiac valve insufficiency and congenital diaphragmatic hernia. The possibly dynamically evolving state of the patient is observed via clinical signs and monitored variables. Clinical signs, such as a palpable pulse, can be evaluated without dedicated medical devices. Monitored variables, such as an invasive blood pressure, do require medical devices for their measurement. Clinical signs and monitored variables that need to be simulated constitute the output variables of the model. Qualitative choices that influence simulation dynamics are those of critical incidents and interventions to be simulated. Examples of critical incidents are acute blood loss and pneumothorax. Interventions such as exercise or the Valsalva maneuver may be carried out by a healthy volunteer. Therapeutic interventions such as chest compression or provision of additional oxygen can be carried out by a health care provider on patients. Some interventions require medical devices, others don't. An overall block diagram with model input (independent) and model output (dependent) variables can now be established (see Fig. 1.9 for an example). If representation of internal structures and functions is not required, and if sufficient input–output data are available, then a black-box model may suffice. In its simplest form this model may take the form of a database or a set of look-up tables for the values of the dependent variables. They may also take the form of difference equations or neural networks with parameters that are estimated based on input–output data (Ljung, 1999). Note that—in principle—detailed physiological knowledge is not required to establish a black-box model. In the remainder of this book we will focus on white-box and gray-box models. For example, it is difficult to imagine that lung perfusion and ductus arteriosus flow rate (Fig. 1.9) can be modeled as a function of all possible combinations of the continuous input variables, merely based on input–output data and without representing internal structures and functions. Therefore, we used a gray-box model. All previous choices may influence the internal structures and functions to be incorporated. For example, to represent fetal hemodynamics, specific structures that are not part of the postpartum circulation, such as the umbilical cord and placenta, need to be included. Qualitative models requirements are summarized in Table 2.1.

Many modeling and simulation projects require only setting a subset of the requirements listed in Table 2.1. For example, not all projects involve pathologies or critical incidents. Conversely, requirements may have to be combined, for example, a critical incident occurring in a patient with an underlying pathology. We repeat the comment made in the introduction to this chapter that model requirements may evolve depending on the phase of a

1. Organ or physiological system
2. Population
3. Physiological states
4. Pathologies
5. Clinical signs and monitored variables
6. Critical incidents
7. Interventions
8. Overall block diagram
9. Internal structures and functions (if applicable)

TABLE 2.1 Qualitative Model Requirements

modeling and simulation project. Given that these requirements guide all subsequent steps of the modeling and simulation process, they should be set with utmost care; adding a critical qualitative requirement toward the end of the process may lead to invalidation of much previous work.

The review problem at the end of this chapter provides detailed examples of questions involved in specifying model requirements, setting the stage for the first review problem in each of the subsequent chapters of Part I of the book.

Based on the qualitative requirements we can continue the modeling and simulation process (see Fig. 2.1) by designing a corresponding conceptual model (Chap. 3). Alternatively, we can complete the requirements as outlined in the subsequent sections, before setting up a conceptual model. In our experience this second approach provides more structure to the overall process.

2.2 Quantitative Aspects

In this section, we elaborate on one of the main qualitative requirements—the model block diagram—and specify quantitative aspects of input and output variables. These requirements and observations will play a role in interface design and model code implementation (Sec. 2.3 and Chap. 5) and in mathematical modeling (Chap. 4). We also consider quantitative aspects of system behavior that have to be included in the mathematical model. For the model input and output variables we:

- Select a symbol, also see Sec. 1.5.
- Assign a unit and use it to specify typical ranges and resolutions.
- Specify bandwidths.

Table 2.2 describes the input variables of the model of Fig. 1.9 in more detail.

The resistances of the umbilical cord and ductus arteriosus (DA) go to infinity on simulation of cord clamping and progressive closure of the DA respectively. For that reason, we use conductance, the inverse of resistance, instead. The electrical engineering symbol for conductance is G. Because these are input variables, rather than parameters, we use $g(t)$. There are no such constraints on the pulmonary vascular resistance. Subscripts indicate the three anatomical structures. These variables are not routinely measured in clinical practice, so we use units that are convenient for modeling purposes right away. The range for the umbilical cord conductance represents complete closure, and a typical term fetal value, respectively (Sá Couto, 2010). The resolution indicates that we want to

Input variables	Symbol	Unit	Range	Resolution	Bandwidth
Umbilical cord conductance	$g_{uc}(t)$	mm Hg^{-1} ml s^{-1}	0–0.27	0.01	1 Hz
Pulmonary vascular resistance	$r_{pv}(t)$	mm Hg ml^{-1} s	2.0–12.0	0.1	0.1 Hz
Ductus arteriosus conductance	$g_{da}(t)$	mm Hg^{-1} ml s^{-1}	0–8.3	0.1	0.01 Hz

TABLE 2.2 Input Variables of the Model for Educational Simulation of Hemodynamic Transitions at Birth

be able to simulate incomplete and progressive closure of the umbilical cord, for example, via partial compression of the cord by the fetus, or partial clamping by the health care provider. Similar considerations hold for the ranges and resolutions of the other two input values. The bandwidth indicates that the umbilical cord conductance can change rather rapidly, but we don't expect the signal to have any frequency components above 1 Hz. The changes in pulmonary vascular resistance take place over minutes, and the changes in DA resistance over hours and days, which is reflected in decreasing bandwidths.

Depending on the use of the model, it may be convenient to specify input and output variables in clinical units and establish only a consistent set of standard units, and convert variables to them, in the mathematical modeling phase—see Chap. 4.

It will be clear from the above example that the ranges, resolution, and bandwidth are indicative, rather than highly precise values. They play a role in displaying and storing data, and can be further specified if a simulator design requires it. Bandwidth is used here in the sense that the variable or signal has no frequency components outside that frequency range. Theoretically, we can reconstruct a signal from its samples if the sampling frequency is at least twice the bandwidth (in this sense). Note that this value is usually larger than the –3 dB definition used in signal processing. The bandwidth of input and output signals also gives an indication of system dynamics, which is of particular importance if real time performance is required; also see the next section and Chap. 5.

The intended use of the model for hemodynamic transitions at birth is real time and predictive, rather than descriptive. For example, trainees will be manipulating the position of the fetus, influencing umbilical cord conductance, and observing the consequences, in real time. The phenomena to be modeled are mainly continuous processes, so the basic mathematical model will be continuous-time. We have already seen that the simulation requires a MIMO representation of the system (Fig. 1.9). The cardiovascular system contains nonlinearities such as unidirectional cardiac valves. These nonlinearities will most likely become apparent in the relationship between input and output variables. The main hemodynamic transitions at birth are represented via the input variables. For the duration of a delivery simulation exercise, the remaining hemodynamic system can be considered time invariant.

2.3 Implementation and Interfacing

Implementation of physical models, Sec. 1.3, involves electromechanical hardware and engineering design.1 Interfaces to the environment are often an integral part of a physical model. In this book we will focus on mathematical models and implementation in software. These models require interfaces that need to be explicitly designed. We will consider requirements in the categories: model code implementation, interfacing, and general program requirements. In each one of these categories, requirements depend on the intended model application and on the phase of model development and use.

Model Code Implementation

General requirements for code implementation of a mathematical model are that the equations are calculated in the correct order and that a simple, accurate, and efficient numerical integration method is used. Especially when coupling multiple dynamic models, the sequence of equations is critical. In Chap. 5 we will present methods that allow us to meet these requirements. Efficiency of code implementation may be more important in final use than in the design and validation phases, especially if real time or accelerated-time performance is required.

Interfacing

An *interface* is a layer between a model and the environment that processes input data from the environment and presents output data to it. We further distinguish run-time and off-line interfaces. Examples of simple interfaces are keyboard entry of a numerical value, or a graph of simulation results. More elaborate interfaces involve emulation of medical devices, such as a clinical monitor, to enter or display data.

In the model development phase, interfaces should be flexible, but don't have to have a highly standardized form. We want to be

---

[1] Which may include design via relatively low cost and flexible mathematical prototyping.

able to initialize the model, change parameter values, set input variables, display output values in numerical or graphical displays, and so on (Fig. 2.2). For model validation, including documentation and scientific publication, data should be presented more carefully, and an appropriate corresponding interface, including scales and units that match the target data, should be chosen (Fig. 2.3). This can be accomplished as an extension to the model development program or off-line, based on saved data and using a specific graphing program.

In a research application or in a model development phase, it may be perfectly acceptable, and even desirable, to "administer" fluid via the keyboard, have the model execute 10 times faster than real time, and present effects, for example, on blood pressure and heart rate, as a graph at the end of a simulation run. In an immersive acute care training simulation, however, the resulting differences with clinical reality would greatly distract the trainees, and thereby most likely reduce both the quality of learning and transfer of learned skills from the simulated to the real environment. In such an application, we would want to use real fluids, and automatically detect their administration, run the model in real time, and present, for example, heart rate continuously as a palpable pulse on a mannequin (Fig. 2.4).

The success of such modeling and simulation projects often stands or falls based on the quality of the interfaces. In the remainder of this book, we will focus on software implementation of

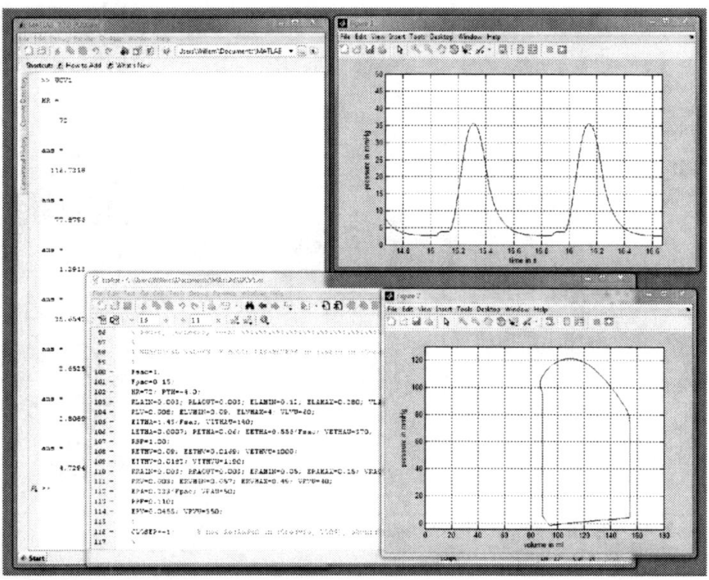

FIGURE 2.2 Model development interface.

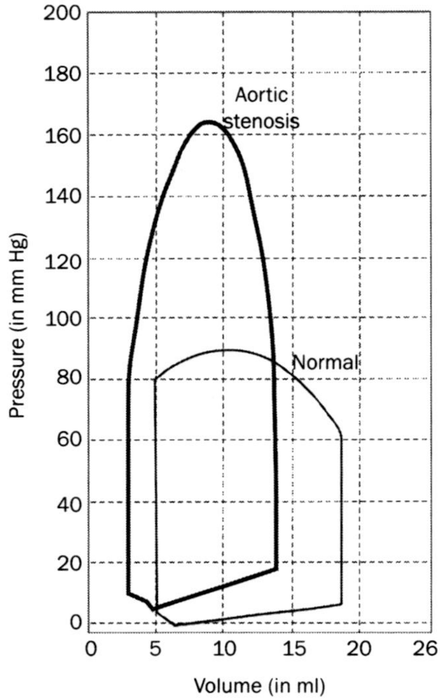

FIGURE 2.3 Model validation interface. (*Adapted from Goodwin, 2004.*)

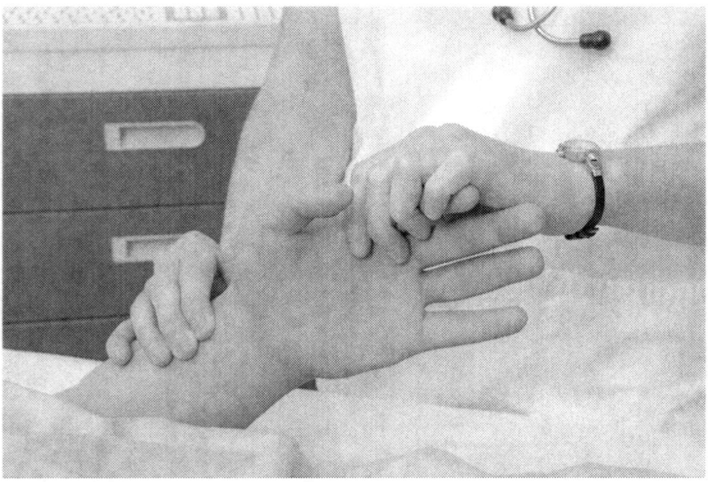

FIGURE 2.4 Palpable pulse for acute care simulation. (*Photo courtesy of Medical Education Technologies, Inc.*)

mathematical models. Only in Chap. 12 will we briefly return to interface design.

General Program Requirements

General program requirements involve the choice of programming language, the need for real-time or accelerated-time ("fast-forward") simulations, and specification of the global program flow. In our opinion, an environment like Matlab (The MathWorks) is a good tool for mathematical model design and validation because of the relatively simple, direct implementation of equations, and its interfacing capabilities. In the design phase, real time behavior is often not a requirement, and for demonstrations approximate real time behavior can easily be accomplished; see the next chapter. A potential shortcoming is that Matlab allows for programming without much preliminary reflection on an algorithm. In our experience, the explicit formulation of an algorithm (see Chap. 5) facilitates debugging, documentation, reuse, and sharing and transferring of developed code. Visually oriented modeling and simulation tools allow for rapid prototyping, but address the conceptual and mathematical modeling and code implementation phases all at once, which can be confusing, and can jeopardize reuse and transfer of code. If a model is incorporated in a simulator that involves other functional units, for example, other models, user interfaces, communication with hardware, then a more general high-level programming language such as Java or C needs to be used. The model algorithm needs to fit the overall program flow. True real-time programming, if required, is a highly specialized task for professional software engineers. The modeler needs to verify that real time is accomplished, and pay close attention to sequencing of model equations, especially if closed-loop systems, which are potentially sensitive to oscillations, are involved.

2.4 Target Response Data

Target data should be established for baseline simulation results and for dynamic responses to changes in input variables. They may take the form of tables or graphs of numerical data, or waveforms. These data should not be used for parameter estimation, part of establishing the mathematical model, also see Chap. 4. The target data should match the modeled population, pathologies, and conditions as closely as possible (also see Sec. 2.1). Population sizes, standard deviations, and ranges should be reported whenever available. Table 2.3 shows an example of target data for fetal hemodynamics from (Treitel, 1987) as used in (Sá Couto, 2010).[2]

[2] In Chap. 6 we will come back to the use of target data from animal experiments.

Anatomical structure	Flow rate, in % of combined ventricular output
Right ventricle	66 ± 18
Left ventricle	34 ± 11
Ductus arteriosus	≈ 57
Pulmonary circulation	8 ± 5
Foramen ovale	26 ± 12
Placenta	35 ± 12

Note: Values expressed as mean ± standard deviation, measurements in 16 term fetal lambs.

TABLE 2.3 Example of Target Data

There are advantages to establishing target data at the onset, rather than during or toward the end of a modeling and simulation project; they provide concrete examples for how the model should respond in a few specific cases. Additional target data may be needed when expanding the range of validity of the models, also see Chap. 6.

Review Problems

2.1 Clinical instructors want to develop a simulation-based class for teaching hemodynamic[3] principles and monitoring. Their request is for a simulator of acute hypotension (low arterial blood pressure) and tachycardia (high heart rate) due to three different possible causes: (1) Low blood volume, (2) decreased systemic vascular resistance, and (3) low myocardial contractility (a weak heart). The goal is for trainees to interact with the simulator and make a differential diagnosis of the underlying cause. To be able to do so, they need access to other monitored hemodynamic variables besides arterial blood pressure and heart rate.

- Refresh your knowledge of relevant physiology. Gather background information on how central venous pressure and cardiac output are measured and on how the three causes for hypotension influence these hemodynamic variables.

The clinical instructors further specify that they would like to manipulate the severity of the underlying causes of hypotension, and that these causes can appear in combination. The simulated arterial blood pressure signal should be available as a noninvasive reading of systolic and diastolic blood

[3] *Hemodynamic* is defined as pertaining to the fluid dynamics of blood in the cardiovascular system.

pressures, and on request, take the form of a pulsatile continuous-time tracing corresponding to an invasive measurement.

- Select model input–output variables and establish a block diagram.

It is unlikely that there are enough input–output data on human subjects to establish a black-box model for this MIMO system. Moreover, manipulation of the model to reflect different clinical conditions and possibly different patients is almost impossible in a black-box model, which has no internal structure reflecting anatomy and physiology.

- Which structures and functions need to be incorporated in the hypotension model—atria, ventricles, systemic circulation, pulmonary circulation, baroreflex, chemoreflex, or other?

The instructors would like to be able to simulate other patients and conditions in the future, but for now are satisfied with a single, representative population.

- Specify the population, taking into considerations educational objectives, for example, a typical hospital population, and access to data to estimate model parameters and validate its response.

Other model requirements will be addressed in subsequent review problems.

References and Further Reading

Berne RM, Levy MN, Koeppen BM, Stanton BA, eds., *Physiology*. 5th ed. Philadelphia: Mosby; 2004.

Goodwin JA, van Meurs WL, Sá Couto CD, Beneken JEW, Graves SA. A model for educational simulation of infant cardiovascular physiology. *Anest Analg* 2004 Dec; 99(6):1655–64.

Guyton AC, Hall JE. *Textbook of medical physiology*. 11th ed. Philadelphia: W.B. Saunders; 2006.

Ljung L. *System identification, theory for the user*, 2nd ed. Upper Saddle River, NJ: Prentice Hall; 1999.

Sá Couto CD, Andriessen P, van Meurs WL, Sá Couto PM, Ayres-de-Campos D. A model for educational simulation of hemodynamic transitions at birth. *Pediatr Res*. 2010; 67(2): 158–65.

Teitel DF, Iwamoto HS, Rudolph AM. Effects of birth-related events on central blood flow patterns. *Pediatr Res*. 1987; 22:557–66.

CHAPTER 3

Conceptual Models

If we need a model with an internal structure and processes we can understand, manipulate, and possibly expand, then it is convenient to specify these qualitative aspects first, before we formalize them and add quantitative aspects in a mathematical model. This step in the modeling process is accomplished via a conceptual model. In this chapter we introduce two predominantly visual representations of internal system structure and function: block diagrams and component diagrams.

In the previous chapter, we specified what we require from a model in terms of input–output behavior and in terms of structures and processes that need to be included. Our goal is to get to a mathematical model that meets these requirements, which can then be transformed into computer code and used to obtain simulation results. We can now carry out a *literature search* to see if an existing model meets all our requirements. If this is the case, we can implement a mathematical model in code or a physical model in hardware. In many cases, however, published models only partially meet our requirements. In such cases they need to be adapted or combined. If no model can be found, a completely new model needs to be designed. If we are satisfied with an empirical model, then we can choose the type and order of a mathematical model, for example, a second-order linear ordinary differential equation or a neural network, and parameter estimation completes the modeling process. Note that sufficient input–output data should be available to estimate the parameters of such a model. If, on the contrary, we need an explanatory model we need to set up a *conceptual model* that reflects the internal structure of the system in terms of subsystems and networks of idealized system components. That is relatively straightforward for manmade engineering systems, which are often designed that way. Physiological systems, however, frequently do not have clearly distinguishable subsystems, and do not consist of close-to-ideal components in a simple network. Yet, engineering-type analytical techniques can be used as an intermediate step between model requirements and explanatory mathematical models. The involved simplifications constitute a delicate balancing act: too few simplifications, and the model may take too long to establish, it may be difficult to maintain, and its software implementation may run too slow for the envisioned application; too many simplifications, and the model may not have a realistic response in all required situations.

3.1 Block Diagrams

Let us assume we want to set up a basic model of the normal cardiorespiratory system. Following the steps listed in Table 2.1, we turn to the standard physiology text books for background information on the cardiorespiratory system. We specifically want to study the effect of oxygen metabolism in body tissues on spontaneous breathing and systemic arterial and venous blood gases, leading to the input and output variables reflected in the block diagram (Fig. 3.1).

We further specify that the model should include spontaneously breathing lungs, distribution of respiratory gases by the circulation, metabolism in body tissues, and control of breathing, which maintains arterial blood gases at appropriate values by adjusting respiratory muscle activity. An intuitive graphical representation of these structures and processes is included in Fig. 3.2.

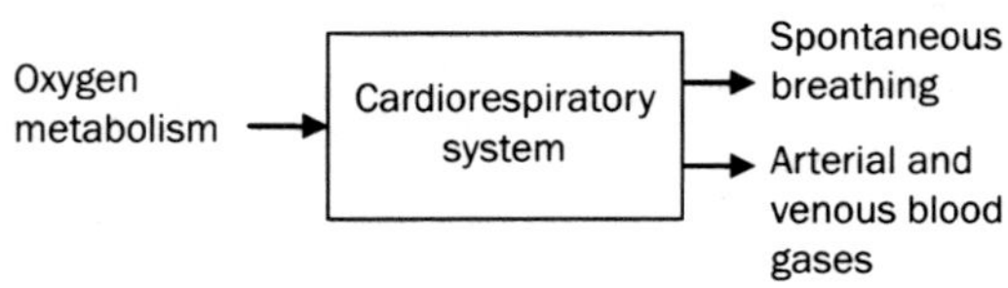

FIGURE 3.1 Block diagram for a basic model of the cardiorespiratory system.

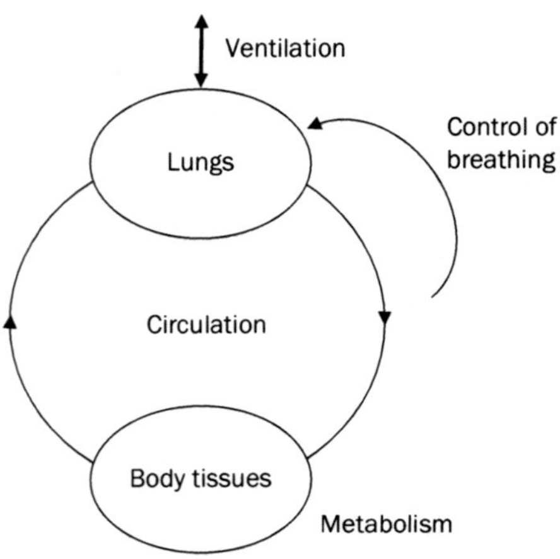

FIGURE 3.2 Graphical representation of the cardiorespiratory system.

A final qualitative requirement is to represent the cardiorespiratory system of an average young, healthy, adult female—a population for which we have explicit target data.

To get to a mathematical model, we need to separate structures and processes more carefully than in Fig. 3.2. The lungs and upper airway are anatomically separate from the overall system. Blood reaches the lungs via the pulmonary arteries and leaves via the pulmonary veins. The partial pressures of the respiratory gases oxygen and carbon dioxide in those vessels are $p_{pa}O_2(t)$, $p_{pa}CO_2(t)$, $p_{pv}O_2(t)$, and $p_{pv}CO_2(t)$, respectively. These are not quite our intended output variables (Fig. 3.1), but in normal circulation the partial pressures of respiratory gases in the pulmonary arteries are very close to those in systemic veins. Similarly, the partial pressures of respiratory gases in the pulmonary veins have values that are very close to those in systemic arteries. The input variable, oxygen metabolism (Fig. 3.1), is represented by $mO_2(t)$. Let us further assume that the respiratory muscles responsible for spontaneous breathing receive nervous activity with a firing frequency $f_{mus}(t)$. Breathing and ventilation are reflected in the time-varying alveolar volume $v_A(t)$, another output variable in Fig. 3.1. The *block diagram* in Fig. 3.3 divides this system into subsystems, and makes key variables and causal interactions between subsystems explicit. Note that control of breathing is not yet included.

Block diagrams are the basis of a *modular approach* to modeling and simulation. We can now focus our attention on a single block, or a block can be reused in another model. In Fig. 3.3 we used a structural or anatomical criterion for separating the system into subsystems. Anatomical hierarchies (see Fig. 3.4 for an example) often provide a first basis for identifying subsystems.

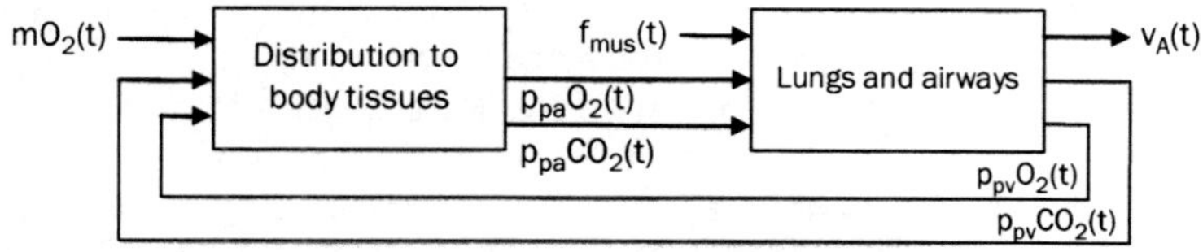

FIGURE 3.3 Block diagram of the cardiorespiratory system.

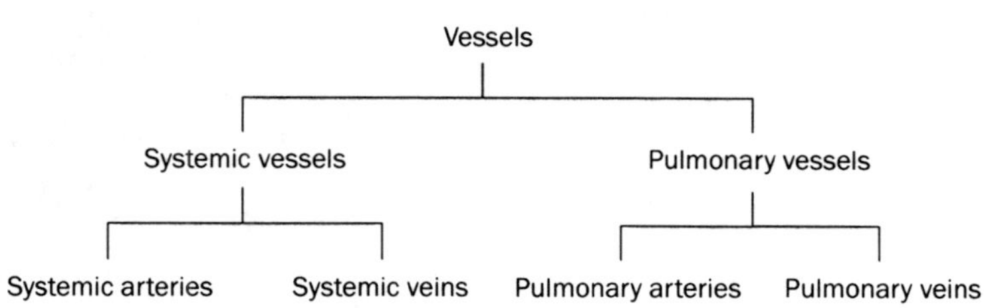

FIGURE 3.4 Basic hierarchy of blood vessels.

A second criterion for separating systems into subsystems is based on function, rather than structure, or on physiology rather than anatomy. Examples of physiological processes include formation, transport, transformation, and absorption of substances and generation, transmission, and processing of signals. The physiological criterion refers to the type of quantity involved in the modeled process. In an adaptation of work presented by Rideout (1991), we distinguish the following *basic types of quantities*:

1. *Volume*. Processes: bulk mass transport, formation and absorption of substances. Examples: respiratory gas flow, blood flow, cerebrospinal fluid production and absorption.
2. *Concentration*. Processes: dissolved mass (solute) transport in a fluid carrier or via diffusion, transformation of substances. Examples: distribution and metabolism of respiratory gases, hormones, and drugs in blood or plasma, heat (under certain conditions).
3. *Action potential*. Processes: transmission and processing of predominantly electrical signals by a fixed carrier. Examples: electrical activity of the heart, transmission of baroreceptor signals via the vagus nerve, and processing of those signals by the brain stem.

A few comments on this classification: transport of respiratory or anesthetic gases in a gas mixture is an intermediate case between bulk and dissolved mass transport; in this case the contribution of solute volume to carrier volume cannot be ignored. Partial pressure is a concept (and variable) that is frequently used alongside concentration in modeling of solute transport. "Information" can be carried by both solutes and signals. At the microscopic level, transmission and processing of signals involve movement of neurotransmitters and electrical charge in the form of electrons or ions.

The lungs in Fig. 3.3 represent a single anatomical structure, but can be further divided into lung mechanics involving volume transport, and pulmonary gas exchange involving solute transport in the volume carrier and diffusion between gas and blood (Fig. 3.5). The main factors that determine the partial pressures of gases in the pulmonary veins are the supply via the pulmonary arteries and ventilation reflected in the time varying alveolar volume $v_A(t)$.[1]

Anatomical and physiological criteria concur to separate the cardiorespiratory system from control influences. The partial pressure of carbon dioxide in systemic arterial blood $p_aCO_2(t)$ plays a primary role in the control of breathing by the autonomic nervous system. Respiratory centers in the brain stem are sensitive to $p_aCO_2(t)$,

[1] Composition of inspired gases is considered constant, for now.

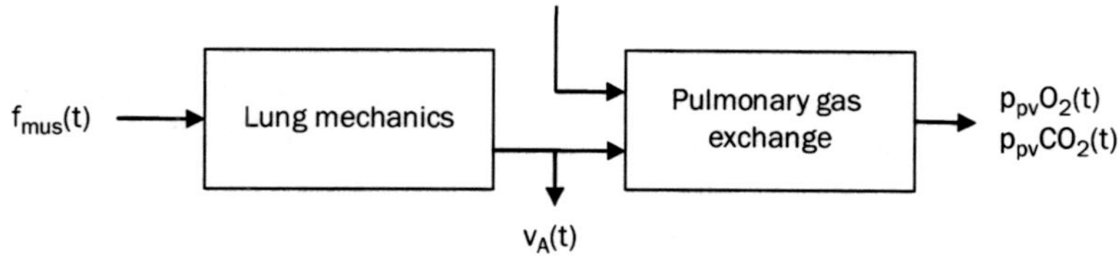

FIGURE 3.5 Block diagram of the lungs and airways. To simplify the representation, a single arrow now represents partial pressures of both gases.

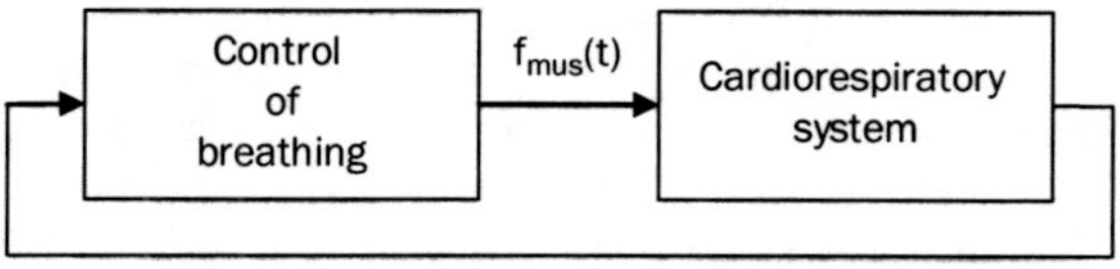

FIGURE 3.6 Block diagram of the cardiorespiratory system and control of breathing.

transforming the partial pressure into time-varying nervous activity sent to the respiratory muscles, the above-mentioned $f_{mus}(t)$ (Fig. 3.6).

A third criterion for separating systems into subsystems is measurability of input and output variables. Measurability facilitates interpretation and system identification, but meaningful subsystems may sometimes rely on input–output variables that are difficult or impossible to measure directly. Most of the variables in the presented example can be measured directly or indirectly. We already observed that, in a normal circulation, the partial pressures in the pulmonary arteries are very close to those in easily accessible systemic veins. The partial pressures in the pulmonary veins are very close to those in systemic arteries, or can be approximated by partial pressures in alveolar or exhaled gases, with which they are in equilibrium. The nervous activity stimulating respiratory muscles could be measured at the phrenic nerve providing motor supply to the diaphragm muscle, but it may be easier to obtain derived signals such as an electromyogram of certain respiratory muscles.

In summary, we have used criteria to separate systems into subsystems based on:

- Anatomy (structure), involving anatomical hierarchies.
- Physiology (function), involving the basic type of quantities: volume, concentration, and action potential.
- Measurability of input and output variables.

The notions of system and subsystem naturally expand into a *hierarchy* of superiors and subordinates. When applied to block diagrams, input–output configurations have to match across levels. For example, the block diagrams of Figs. 3.3, 3.5, and 3.6 fit the hierarchy shown in Fig. 3.7.

In many modeling and simulation projects, a block diagram hierarchy will be established in a "top-down" approach: starting from the single overall block part of the requirements, we work our way down until it is no longer necessary or practical to distinguish further subsystems. At the bottom of the hierarchy there may be a simple equation describing the relationship between input and output variables, or we may choose to include a black box for the particular subsystem, and estimate parameters based on input–output data. A third possibility is that the subsystem can be further described in terms of a network of interacting components reflecting basic system properties. We will explore this form of conceptual modeling in the next section. For relatively complex conceptual models, maintaining the hierarchy may be useful to keep oversight. For the relatively simple model of Fig. 3.7, after establishing the hierarchy as part of the conceptual modeling process, we can substitute the lower blocks into the superior levels, leading to the more compact representation of Fig. 3.8.

Note how the overall system is divided into subsystems along anatomical lines and according to type of quantities. Comparing Figs. 3.8 and 3.1, we have indeed formalized a lot of qualitative

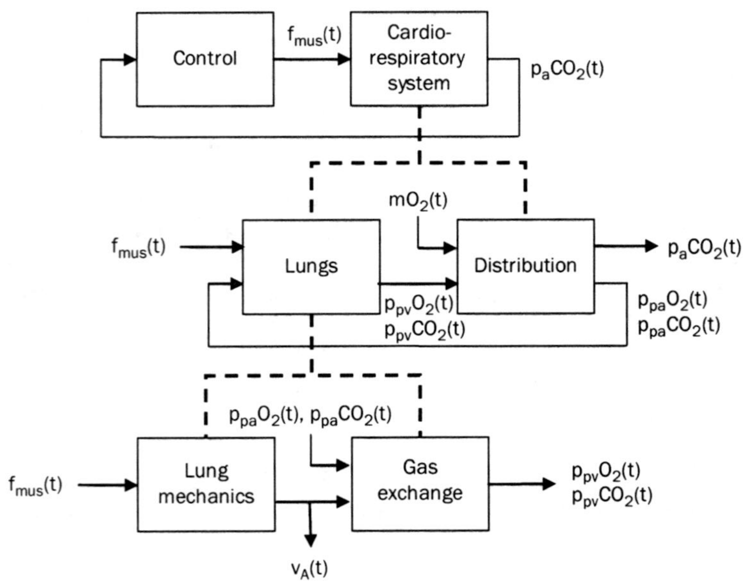

FIGURE 3.7 Hierarchy of block diagrams of the cardiorespiratory system. The blocks of Fig. 3.3 are re-arranged to highlight the fit with the superior level.

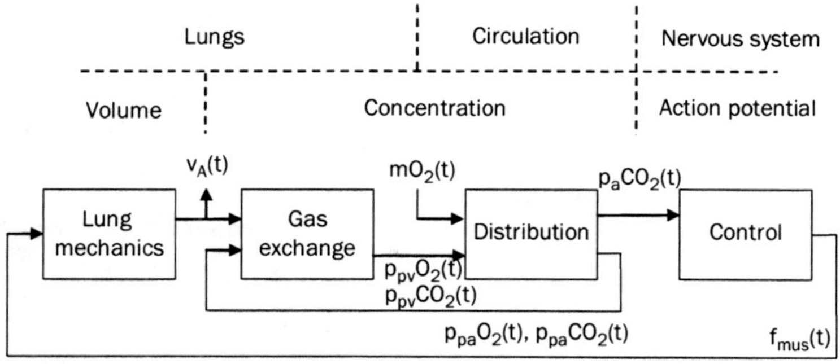

FIGURE 3.8 Integrated (single-level) block diagram of the cardiorespiratory system. Anatomical structures and types of quantities involved in the subsystems are shown above the blocks.

information about the system. For some modeling tasks the "hierarchy" may only be one level deep, in which case we do not distinguish subsystems, and immediately formulate mathematical models or use conceptual models in the form of component diagrams. In large and relatively unconstrained modeling projects, the process of establishing a block diagram hierarchy may be turned around, and anatomy and types of quantities may be used to build a model from the "bottom-up," taking measurability of variables into account only at higher levels.

Note that the block diagrams in this section are presented to illustrate concepts; they contain many simplifications, most notably concerning the limited representation of the circulation and its control, which will be covered in more detail in Part II.

3.2 Component Diagrams

Even if meaningful causally connected subsystems can no longer be distinguished, we may be able to describe the internal structure and functioning of a system further in terms of interacting *components* connected in a *network*. Overall system behavior depends on the properties of the components and on how they are connected. Figure 3.9 gives an example of a component diagram for an electrical circuit.

The components in this diagram are an external voltage source, a capacitor, a resistor, and an inductor. These may be actual physical components, or represent generalized system properties, for example, when modeling the resonant properties of a cable. In the next chapter, we will see how a mathematical model describing, for example, the relationship between the variables voltage $e(t)$ and charge on the capacitor $q(t)$ can be derived from component descriptions and the laws governing connection of components. Generally speaking, physiological systems do not consist of discrete components connected in a network. However, component diagrams and derived mathematical models may represent system properties and behavior in a reasonable approximation. Figure 3.10 shows examples of components and networks for systems involving the different types of quantities introduced in the previous section.

Setting up a conceptual model of a physiological system in the form of a component diagram is an inductive, creative process, where

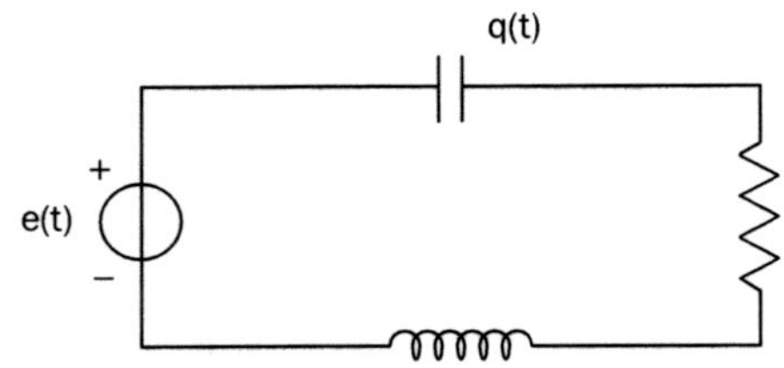

Figure 3.9 Electrical circuit diagram. See the main text for a description of variables and components.

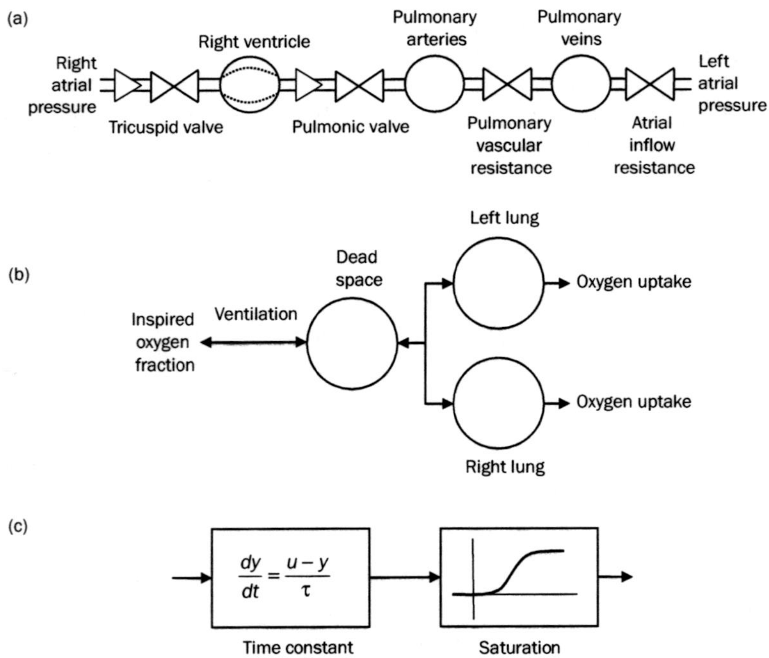

Figure 3.10 Conceptual models of physiological systems in the form of component diagrams: (a) Volume: fluid circuit diagram of the right ventricle and pulmonary circulation. (b) Concentration: compartment model of oxygen mixing in the lungs. (c) Action potential: elementary transfers in the sympathetic nervous system.

simplifying assumptions need to be balanced against anticipated model performance. For example, the succession of tricuspid valve, right ventricle, and pulmonic valve in Fig. 3.10(a) closely reflects basic anatomy and will be appropriate for a broad range of model applications. However, representing the whole pulmonary vascular bed by only three hydraulic components may be appropriate for a model representing the basic relationships between volume, pressure, and flow rate, but not for a model for the study of blood flow distribution in the lungs. The component diagram (of Fig. 3.10(b)) reflects the assumption that oxygen exchange in the lungs can be represented by three distinct spaces with a homogeneous gas concentration in each space, two with oxygen uptake, and one dead space compartment without oxygen exchange. This particular conceptual model further reflects that ventilation of these spaces is bidirectional and that the parallel lung compartments are connected in series with the dead space compartment. Note that for a number of applications, a model based on continuous, unidirectional ventilation of a single homogenous space (Fig. 3.11; also see Riley, 1949) is sufficiently accurate.

If we split the oxygen uptake in the model of Fig. 3.10(b) into uptake by the arterial blood and delivery to the lungs by the venous blood, it becomes a conceptual model for oxygen exchange, part of the gas exchange block of Fig. 3.8, illustrating how block and component diagrams can be used in conjunction. The transfers of Fig. 3.10(c) reflect elementary mathematical operations. These operations are combined in a block diagram.

Conceptual models in the form of component diagrams take on their full meaning in conjunction with the quantitative relationships between the involved variables and the laws governing component connections. These will be introduced formally and systematically in the next chapter, providing a new perspective on what may now seem like rather intuitive, but somewhat ambiguous drawings. In Chap. 4 we will also make the distinction between component diagrams in the same physical domain as the modeled process, and

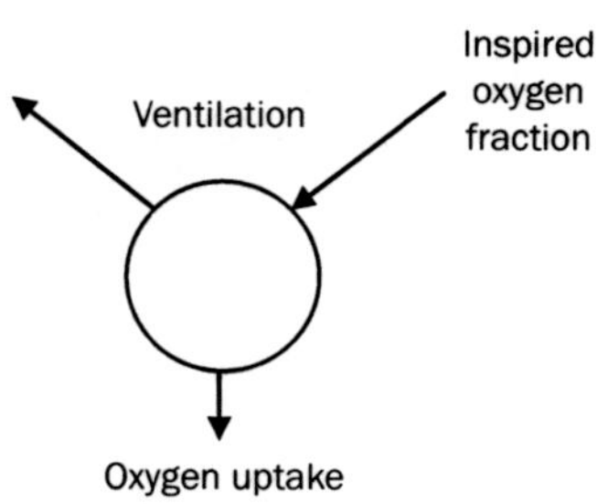

FIGURE 3.11 Simplified oxygen exchange model.

analog representations in another physical domain. We will develop the fluid volume–electrical charge analogy, which implies that network techniques developed in electrical engineering can also be applied to modeling fluid mechanics. Part II of the book provides further detailed examples of conceptual and mathematical model design.

3.3 General Observations on Conceptual Models

The traditional engineering disciplines have well-established standard symbols for component diagrams. Wherever appropriate, we will use those. However, in modeling of physiological systems, we often encounter a number of different physicochemical systems simultaneously, and in the scientific literature a number of rather loose conventions are used. Our choice of symbols is a personal one, as much as possible inspired by standard symbols, but with the additional requirement of maintaining the distinction between systems with different basic types of variables. An example of such a choice is the use of circular rather than rectangular representations of compartments, to avoid mistaking them for block diagrams and misinterpreting represented connections: an arrow in a block diagram represents a causal connection, but in a compartment model it represents positive flow of fluid and solute, or metabolism. The (nonstandard) use of parallel lines for flow conduction in fluid circuit diagrams such as Fig. 3.10(a) results from similar considerations.

Block diagrams and component diagrams are powerful complementary conceptual modeling tools for continuous-time dynamic systems, but not the only ones. Any representation that links system structure and processes to a mathematical model, and that is convenient for both the modeler and model user, can be applied for conceptual modeling and more general discussions and presentations of the model.

In modeling of physiological systems, conceptual modeling is a step in which a lot of domain knowledge (anatomy, physiology, physics, etc.) and several critical assumptions are incorporated. Typically, it is the step in which the dialog between life scientists and modelers, which started with transfer of background knowledge and requirement formulation, is most pronounced. Simplifications at this stage have consequences in all subsequent phases, but may only become apparent in the final stages of development or even during practical use of a model. This constitutes one of the major challenges in physiologic modeling and simulation. A common vocabulary and a combination of knowledge and experience will guide the modeler and model user in distinguishing useful simplifications from detrimental oversimplifications. A rigorously documented modeling process allows revisiting simplifications, if necessary.

Review Problems

3.1 Consider the requirements formulated as part of review problem 2.1. Divide the overall system into two subsystems, one for the baroreflex and the other for the "uncontrolled cardiovascular system." Carefully identify the input and output variables of each subsystem. Which criteria did you use?

3.2 Consider the electromechanical properties of the heart and the transmission of electrical activity to the chest wall, where it results in polarity changes that can be seen in an electrocardiogram (ECG). Design a conceptual model in the form of a block diagram that separates the intrinsic electrical activity of the heart from the transmission to the chest wall and from the mechanical activity resulting in a time-varying contraction. What are the key variables?

3.3 How does the block diagram from Problem 3.2 need to be modified to represent mechanical and electrical activity involved in third-degree heart block, which is a disorder of the cardiac conduction system where there is no conduction through the atrioventricular node and complete dissociation of the atrial and ventricular activity, the ventricles beating at their intrinsic rate of approximately 40 beats per minute?

References and Further Reading

Berne RM, Levy MN, Koeppen BM, Stanton BA, eds. *Physiology*. 5th ed. Philadelphia: Mosby; 2004.

Guyton AC, Hall JE. *Textbook of medical physiology*. 11th ed. Philadelphia: W.B. Saunders; 2006.

Rideout VC. *Mathematical and computer modeling of physiological systems*. Madison, Prentice Hall, 1991.

Riley RL, Cournand A. 'Ideal' alveolar air and the analysis of ventilation-perfusion relationships in the lungs. *J Appl Physiol*. Jul 1949; 1(12):825–847.

CHAPTER 4

Mathematical Models

In this chapter we present a method for modeling causal, continuous-time, finite-dimensional, dynamic systems. A mathematical model, corresponding to two conceptual models in the form of component diagrams, is derived based on component descriptions and natural or empirical laws. The state variable representation, in its most general formulation, captures the behavior of SISO and MIMO systems, and of static and dynamic, linear and nonlinear, and time-invariant and time-variant systems. Numerical integration and simulation of state variable models is straightforward, and state variable models can be derived from conceptual models of biomedical systems in the form of component diagrams. After a brief reminder of numerical values and units, we present fluid circuit elements that can be used to model circulation and ventilation; homogeneous compartments that can be used to model solute exchange, transport, and metabolism in the respiratory system; and elementary transfers that can be used to model the nervous system. We introduce the analog and its applications, and elaborate on the analogy between fluid and electrical circuits, presenting network reduction techniques for parallel and in series elements, and an extensive example of deriving a state variable model of fluid circuits.

4.1 A Model of Two Physical Systems

In this section we will see how a mathematical model is derived from a conceptual model in the form of a component diagram. This step involves formulating and combining natural or empirical laws describing components and connections. To highlight the general applicability of modeling and simulation, we introduce this step with two examples from mechanical and electrical engineering, respectively. Mathematical models for specific biomedical systems will be introduced in the second half of this chapter. Consider the translational mechanical system of Fig. 4.1.

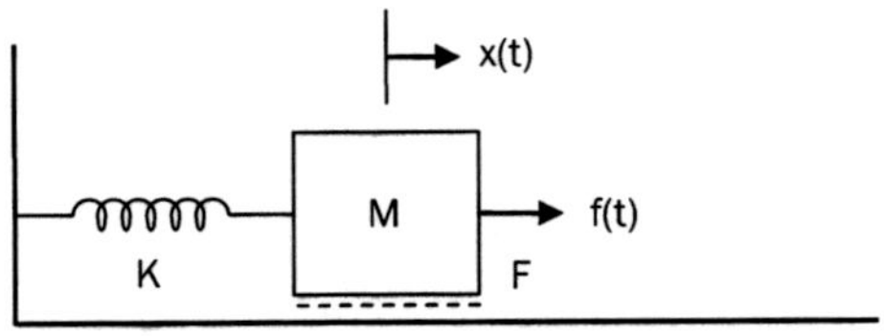

FIGURE 4.1 Translational mechanical system. See the main text for a description of components, variables, and parameters.

The components in this diagram are a spring, a mass, and a frictional element. Their behavior can be depicted in terms of relationships between variables such as forces and displacements. For example, Fig. 4.2 shows the relationship between the force on a spring and its deflection.

These relationships can often be described by equations. Over a limited range of the variables, a linear equation characterized by a single constant or parameter may be an acceptable approximation. For example, let $f_s(t)$ be the force exerted by the spring on the mass. For small deviations, this force and the extension $x(t)$ are proportional.

$$f_s(t) = -Kx(t) \tag{4.1}$$

where K is the spring constant. The minus sign indicates that the force operates in the opposite direction of positive $x(t)$. Let $f_f(t)$ be the force caused by friction of the mass on the surface. If we ignore breakaway forces, necessary to get the object moving, then, in a first approximation, this force is proportional to the velocity $v(t)$ of the frictional element with respect to the surface.

$$f_f(t) = -Fv(t) \tag{4.2}$$

where F is the friction constant. The velocity of the mass is equal to the derivative of its position $x(t)$.

$$f_f(t) = -Fv(t) = -F\frac{dx(t)}{dt} \tag{4.3}$$

The net force on the mass $f_m(t)$ and its acceleration $a(t)$ are linked via Newton's second law.[1]

[1] For more information on this and other natural laws mentioned in this chapter, we point to standard physics texts books (e.g., Cutnell, 1998).

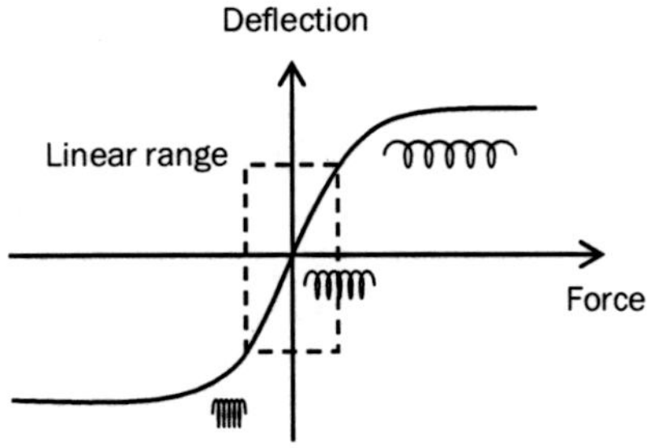

FIGURE 4.2 Spring characteristic.

$$f_m(t) = Ma(t) = M\frac{d^2x(t)}{dt^2} \tag{4.4}$$

Let us further assume that an external force $f(t)$ is applied. The vector sum of the forces on the mass is the sum of the external force and those exerted by friction and spring.

$$f_m(t) = f(t) + f_f(t) + f_s(t) \tag{4.5}$$

Substituting the component descriptions in Eq. (4.5) and rearranging the terms, we find:

$$M\frac{d^2x(t)}{dt^2} + F\frac{dx(t)}{dt} + Kx(t) = f(t) \tag{4.6}$$

Equation (4.6) constitutes a mathematical model of the translational mechanical system represented in Fig. 4.1. This ordinary differential equation (ODE) relates the dependent variable, position $x(t)$, to the independent variable, external force $f(t)$. The other dependent variables velocity and acceleration can be derived from $x(t)$. The specific time profile of $x(t)$ depends on the initial values $x(t_0)$ and $v(t_0)$, on the time profile of the external force $f(t)$ for $t > t_0$, and on the numerical values of the constants or parameters M, F, and K, describing the components. The ODE is a convenient model formulation for mathematical analysis of system behavior. We can derive homogeneous and nonhomogeneous solutions of Eq. (4.6) or, because the coefficients M, F, and K, are constant, use the Laplace transform to compute the transfer function and the nonhomogeneous solution, (see for example Roberts, 2003). In the next section we will present a formulation that is more convenient for numerical integration and simulation. But first consider the electrical system of Fig. 3.9, repeated in Fig. 4.3 below with a number of specific variables and parameters.

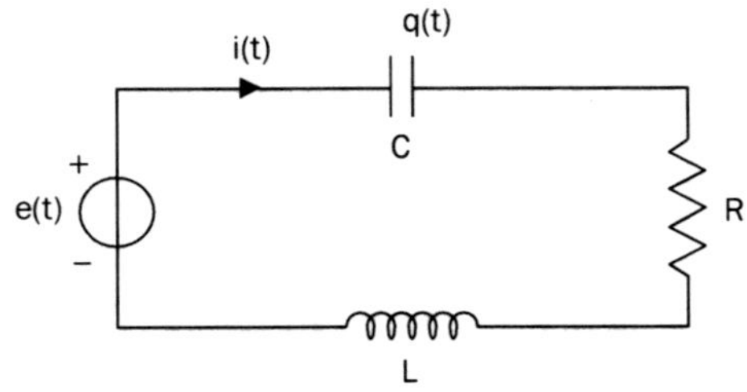

FIGURE 4.3 Electrical system.

The components in this diagram are an external voltage source, a capacitor, a resistor, and an inductor. Their behavior, this time in terms of the relationship between variable charges, voltages, and currents, can also be described by equations. When in a linear form, those relationships can also be characterized by a single constant. For example, let $v_c(t)$ be the voltage across the capacitor and $q(t)$ its charge. Voltage and charge are proportional over a wide range.

$$v_c(t) = \frac{1}{C}q(t) \tag{4.7}$$

where C is the capacitance. The currents through the three components in series $i_c(t)$, $i_r(t)$, and $i_l(t)$, are all the same.

$$i_c(t) = i_r(t) = i_l(t) = i(t) \tag{4.8}$$

Using conservation of charge, we find that this current $i(t)$ is equal to the change in charge on the capacitor.

$$i(t) = \frac{dq(t)}{dt} \tag{4.9}$$

Let $v_r(t)$ be the voltage across the resistor. According to Ohm's law, this voltage is proportional to the current through the resistor $i_r(t)$.

$$v_r(t) = Ri_r(t) = Ri(t) = R\frac{dq(t)}{dt} \tag{4.10}$$

where R is the resistance. The relationship between voltage across $v_l(t)$ and current through an inductor $i_l(t)$ is:

$$v_l(t) = L\frac{di_l(t)}{dt} = L\frac{di(t)}{dt} = L\frac{d^2q(t)}{dt^2} \tag{4.11}$$

Considering the externally applied voltage $e(t)$ and using Kirchhoff's voltage law, which states that the sum of all directed voltages in a closed circuit is zero, we find:

$$e(t) = v_c(t) + v_r(t) + v_l(t) \tag{4.12}$$

and substituting the previously found voltages in terms of charge, we find a mathematical model of the electrical system of Fig. 4.3.

$$L\frac{d^2q(t)}{dt^2} + R\frac{dq(t)}{dt} + \frac{1}{C}q(t) = e(t) \tag{4.13}$$

This ODE relates the dependent variable charge on the capacitor $q(t)$ to the independent variable, the externally applied voltage $e(t)$. Other dependent variables of interest, such as the current $i(t)$ or voltages across the components, can be derived from $q(t)$ using the above equations. The specific time profile of $q(t)$ depends on the initial values $q(t_0)$ and $i(t_0) = dq(t)/dt$ for $t = t_0$, on the time profile of the externally applied voltage $e(t)$ for $t > t_0$, and on the numerical values of parameters R, L, and C. It will not have escaped the attentive reader that Eq. (4.13) is very similar to Eq. (4.6), repeated below.

$$M\frac{d^2x(t)}{dt^2} + F\frac{dx(t)}{dt} + Kx(t) = f(t) \tag{4.14}$$

The two systems are called *analogs*. Despite their different nature, and the different component descriptions and laws used in derivation, their mathematical models reflect a deep similarity in dynamic behavior. Consequences and uses of such similarities will be explored in Sec. 4.7.

Summarizing: Based on a conceptual model in the form of a component diagram, domain specific component descriptions, and natural or empirical laws, we derived mathematical models relating variables of interest. In the next section we will present a model formulation that is applicable to a broad range of systems and more convenient for numerical integration and simulation.

4.2 State Variable Models

In this section we put one of the mathematical models derived in Sec. 4.1 in the form of a set of algebraic equations for first-order derivatives of the so-called state variables. We then generalize these state variable models and show that they can represent a broad class of systems.

An alternative formulation of the mathematical model of the electrical system of Fig. 4.3 can be found by considering the voltage across the capacitor $v_c(t)$ and the current through the inductor $i_l(t)$. Using relationships listed in the previous section, we find their first-order derivatives.

$$\frac{dv_c(t)}{dt} = \frac{1}{C} i_l(t) \tag{4.15}$$

and

$$\frac{di_l(t)}{dt} = -\frac{1}{L} v_c(t) - \frac{R}{L} i_l(t) + \frac{1}{L} e(t) \tag{4.16}$$

Let us assume we are interested in the dependent variable voltage across the inductor.

$$v_l(t) = L\frac{di_l(t)}{dt} = -v_c(t) - Ri_l(t) + e(t) \tag{4.17}$$

Introducing the variable vector:

$$x(t) = \begin{bmatrix} v_c(t) \\ i_l(t) \end{bmatrix} \tag{4.18}$$

and the scalar variables:

$$u(t) = e(t) \tag{4.19}$$

$$y(t) = v_l(t) \tag{4.20}$$

we can write the model of the electrical system, Eqs. (4.15) – (4.17), in the compact form:

$$\frac{dx(t)}{dt} = Ax(t) + bu(t) \tag{4.21}$$

$$y(t) = c^T x(t) + du(t) \tag{4.22}$$

with the constant matrix:

$$A = \begin{bmatrix} 0 & 1/C \\ -1/L & -R/L \end{bmatrix} \tag{4.23}$$

the constant vectors:

$$b = \begin{bmatrix} 0 \\ 1/L \end{bmatrix} \tag{4.24}$$

$$c = \begin{bmatrix} -1 \\ -R \end{bmatrix} \tag{4.25}$$

and the scalar constant $d = 1$. Simplifying the notation of the derivative, and allowing $u(t)$ and $y(t)$ to be vectors, and b, c, and d matrices, Eqs. (4.21) and (4.22) can be generalized to:

$$\dot{x}(t) = Ax(t) + Bu(t) \tag{4.26}$$

$$y(t) = Cx(t) + Du(t) \tag{4.27}$$

Equations (4.26) and (4.27) can be further generalized to:

$$\dot{x}(t) = f(x(t), u(t), t) \tag{4.28}$$

$$y(t) = g(x(t), u(t), t) \tag{4.29}$$

where $f(.)$ and $g(.)$ are algebraic or transcendental functions[2] of the *state variable* $x(t)$, the *input variable* $u(t)$, and time t; $y(t)$ is the *output variable*. Equation (4.28) is called the *state equation*, and Eq. (4.29) is called the *output equation*. For now, we will assume that our models are pure white-box models (Chap. 1) and that the numerical values of *parameters* (the constants that appear in the state and output equations) are given, or can be derived via simple one point measurements, for example, a resistance from pressure and flow data. Let $x(t_0)$ be the initial value of the state variable at time t_0. A block diagram of this model is given in Fig. 4.4.

Let us consider Eq. (4.28) in more detail. If we know $x(t_0)$ and $u(t)$ for $t \geq t_0$, then we can predict the evolution of $x(t)$, and via Eq. (4.29), of $y(t)$. In the next chapter, we will present a practical implementation of this prediction in the form of numerical integration. Numerical integration is facilitated by the fact that the constituent equations of Eq. (4.28) only contain a single, first-order derivative. Because no past

[2] Algebraic functions can be formed by the algebraic operations: addition, multiplication, division, and taking an nth root. A transcendental function cannot be expressed in terms of a finite sequence of these algebraic operations. Examples of transcendental functions include exponential, logarithmic, and trigonometric functions.

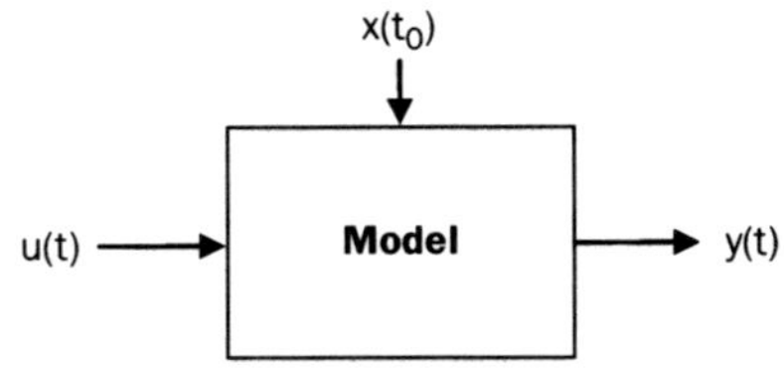

FIGURE 4.4 Block diagram of a state variable model.

information is necessary to predict the evolution of the system, we say that $x(t)$ contains the "memory" of the system.

As state variables we often choose variables that, in physical terms, represent temporary energy storage. This is the case for the voltage across the capacitor and the current through the inductor in the previous section, and in review problem 4.2 we will derive a state variable model for the translational mechanical system of Fig. 4.1, using the extension of the spring $x(t)$, and velocity of the mass $v(t)$ as state variables. The frictional element dissipates mechanical energy, but does not store it. To avoid confusion about notation, we introduce the state variable vector:

$$\overline{x}(t) = \begin{bmatrix} x(t) \\ v(t) \end{bmatrix} \tag{4.30}$$

The evolution of this system depends on the present values of the state variables (extension of the spring and the velocity of the mass) and on the present and future values of the input variable (the external force acting on the system). *How* the present state was achieved has no influence on the evolution of the system. We again find that the memory of the system is stored in the state variables.

To simplify modeling and make numerical integration more efficient, we like to keep the set of state variables as small as possible. This is achieved by selecting *independent state variables*, that is, state variables that cannot be expressed in terms of one or more of the other state variables. For electrical circuits, there are systematic approaches for selecting a set of independent state variables and deriving the corresponding state variable model (Athans, 1974). However, these methods are quite elaborate, and we will see in the subsequent sections and in Part II that in many practical cases in biomedical engineering, independent state variables can be selected, and models derived, by using simple rules of thumb.

We saw in Sec. 1.2 that knowing system properties such as linearity and time invariance is important because certain analysis techniques apply only to, for example, linear, time-invariant (LTI) systems. In the modeling and simulation context, knowing system and model properties also has several practical consequences. For example, if we know the response from zero initial conditions $y(t)$ of a linear system to the input $u(t)$, then we no longer need to derive or compute the zero state response to input $2u(t)$, as, by the system properties, it is simply $2y(t)$. Another example: A static system has no state variables and, therefore, no initial values have to be set to simulate the model. The properties of a model in state variable notation can be recognized by inspection of the equations. Table 4.1 provides the key to such connections. Proof of these associations can be provided based on definitions of system properties, Sec. 2.2. Note that the properties of the model equations do not automatically carry over to the real system it represents. For example, a nonlinear system can have a linear model for a limited range of the input and state variables.

The last property in Table 4.1 is added here for later reference. When notations in Table 4.1 are combined, then so are model properties. For example, when we combine the notations of the third and fourth models in Table 4.1, we obtain Eqs. (4.26) and (4.27), with constant rather than time-dependent matrices A, B, C, and D. Combining the properties, we find that this model is LTI.

In Sec. 4.8 we will come back to some potential concerns with state variables models, but note that the above demonstrates how versatile the state variable representation is: In its most general formulation, Eqs. (4.28) and (4.29), it captures the behavior of SISO and MIMO systems, and of static and dynamic, linear and nonlinear,

$\dot{x}(t) = f(x(t), u(t), t)$ $y(t) = g(x(t), u(t), t)$	$u(t)$ and $y(t)$ scalars	Single-input single-output (SISO)
$y(t) = g(u(t), t)$	no state equation	Static
$\dot{x}(t) = A(t)x(t) + B(t)u(t)$ $y(t) = C(t)x(t) + D(t)u(t)$	$A(t)$, $B(t)$, $C(t)$, and $D(t)$ time-dependent matrices	Linear
$\dot{x}(t) = f(x(t), u(t))$ $y(t) = g(x(t), u(t))$	$f(.)$ and $g(.)$ are not (direct) functions of t	Time-invariant
$\dot{x}(t) = f(x(t), u(t), t)$ $y(t) = g(x(t), t)$		No immediate throughput from input to output

Table 4.1 Equations and Properties of State Variable Models

as well as time-invariant and time-variant systems. In the next chapter, we will see that numerical integration and simulation of state variable models is straightforward. In the remainder of this chapter and in Part II of the book we will see how state variables models can be derived from conceptual models of biomedical systems in the form of component diagrams, but first we provide the reader with a brief reminder of units and numerical values.

4.3 Units and Numerical Values

So far, we have been discussing physical systems and models without referring to specific numerical values. This is a fruitful abstraction, but if we want to obtain specific analytical solutions or numerically integrate the state equation, we need to assign numerical values to the involved quantities: variables, parameters, and the special variable time. This implies the use of a consistent set of units, which is especially challenging in the biomedical field, where a number of nonstandard units are still in use. For smaller modeling and simulation projects, it may be simpler to work with clinical units whenever possible, paying close attention to consistency. For more complex models, and especially when combining models of different systems, we propose to use the SI (from the French *Système International*) units and then convert input and output variables from and to clinically applied units whenever necessary. As a reminder, Tables 4.2, 4.3, and 4.4 list the SI base units, selected SI derived units, and SI prefixes, respectively. The latter help keep numerical values within convenient ranges. Frequently used clinical units for time, volume, and pressure, and their conversions to SI units are listed in Table 4.5. See the literature on physics for the conditions such as pressure and temperature, which are sometimes involved in these conversions. Clinical units for derived quantities, such as flow rate and resistance, will be listed in subsequent sections.

Quantity	Name	Symbol
Length	meter	m
Mass	kilogram	kg
Time	second	s
Temperature	kelvin	K
Amount of substance	mole	mol
Electric current	ampere	A
Luminous intensity	candela	cd

Table 4.2 SI Base Units

Quantity	Name	Symbol and derivation
Area	square meter	m^2
Volume	cubic meter	m^3
Velocity	meter per second	m/s
Acceleration	meter per second squared	m/s^2
Force	newton	$N = kg{\cdot}m/s^2$
Pressure	pascal	$Pa = N/m^2$
Work, Energy	joule	J = N·m
Power	watt	W = J/s
Frequency	hertz	Hz = 1/s
Electric potential	volt	V = W/A
Charge	coulomb	C = A·s
Resistance	ohm	Ω = V/A
Capacitance	farad	F = A·s/V
Inductance	henry	H = V·s/A

Table 4.3 Selected SI Derived Units

Factor	Prefix	Symbol
10^{12}	tera	T
10^{9}	giga	G
10^{6}	mega	M
10^{3}	kilo	k
10^{2}	hecto	h
10	deka	da
10^{-1}	deci	d
10^{-2}	centi	c
10^{-3}	milli	m
10^{-6}	micro	μ
10^{-9}	nano	n
10^{-12}	pico	p

Table 4.4 Prefixes

Quantity	Unit	Symbol and derivation
Time	minute	1 min = 60 s
	hour	1 h = 1 hr = 3600 s
Volume	liter	1 L = 10^{-3} m^3
	milliliter	1 mL (or ml) = 10^{-3} L = 10^{-6} m^3
Pressure	bar	1 bar = 100 kPa
	atmosphere	1 atm ≈ 101.3 kPa
	millimeter mercury	760 mm Hg = 1 atm ≈ 101.3 kPa
	torr	760 torr = 1 atm ≈ 101.3 kPa
	centimeter water	1000 cm H_2O ≈ 98.1 kPa
	dyne per square centimeter	1 dyne/cm^2 = 0.0001 kPa

TABLE 4.5 Frequently Used Clinical Units and Conversions to SI Units

4.4 Direct Representation of Fluid Circuits

We have already seen the basics of mathematical modeling of physical systems: how to go from a conceptual to a mathematical model, the versatile and easy-to-use state variable representation, and numerical values and units. In the remaining sections of this chapter, we will expand those tools for use in modeling of physiological systems. We will call graphical representations such as Figs. 4.1 and 4.3 *direct representations* of the respective mechanical and electrical systems. We will reserve the term *analog* for a graphical representation in a different physical domain from that of the original system. In Sec. 4.7 we will come back to analogs and their use in modeling of physiological systems.

Fluids include both liquids and gases. Basic concepts for the description of the behavior of fluids, such as density, pressure, and continuity, can be found in many introductory texts on physics (e.g., Haliday, 2001). Direct representations of fluid circuit elements that are frequently used in modeling of systems involving volume type quantities such as the circulation and lung mechanics are listed in Table 4.6, together with idealized linear or piecewise linear relationships and mathematical descriptions. Table 4.7 lists the involved basic quantities.

Basic passive fluid circuit elements listed in Table 4.6 include a resistor, compliance, inductor, and valve. The relationship and

Name	Symbol	Relationship	Mathematical description
Resistor	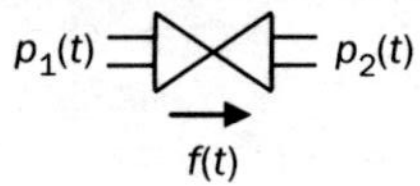	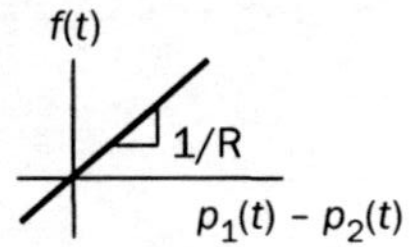	$f(t)=\dfrac{p_1(t)-p_2(t)}{R}$
Compliance	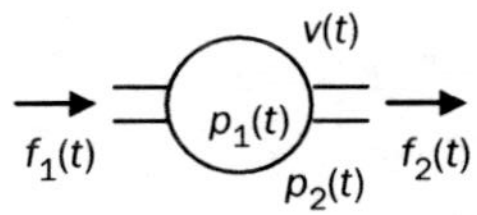	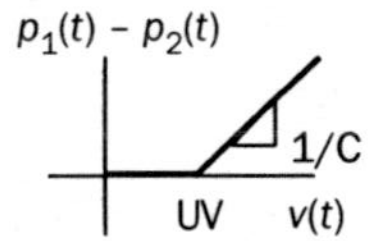	if $v(t) > UV$ $p_1(t)-p_2(t)=\dfrac{v(t)-UV}{C}$ *else* $p_1(t)-p_2(t)=0$ $\dfrac{df(t)}{dt}=f_1(t)-f_2(t)$
Inductor	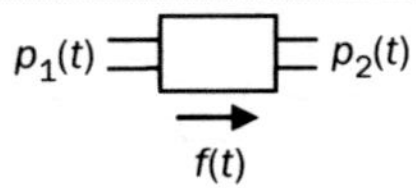	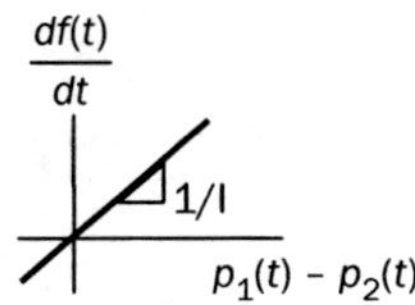	$\dfrac{df(t)}{dt}=\dfrac{p_1(t)-p_2(t)}{I}$
Valve	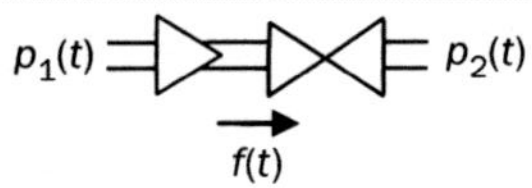	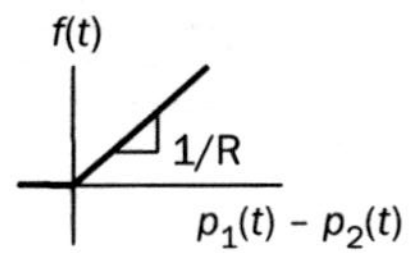	if $p_1(t)-p_2(t)>0$ $f(t)=\dfrac{p_1(t)-p_2(t)}{R}$ else $f(t)=0$

Table 4.6 Direct Representations of Fluid Circuit Elements

Name	Symbol	Relationship	Mathematical description
Compressor	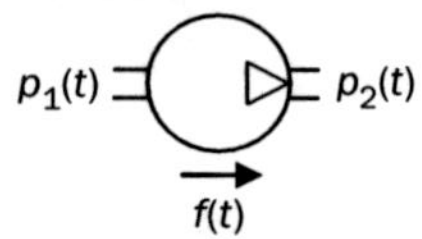	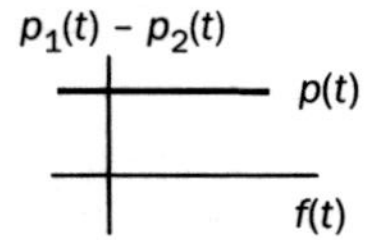	$p_2(t) - p_1(t) = p(t)$
Time-varying compliance	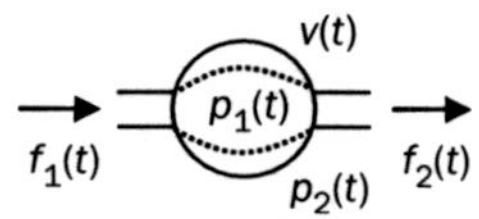	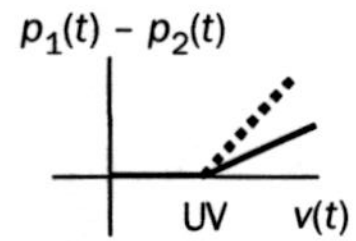	See also compliance $p_1(t) - p_2(t) = \dfrac{v(t) - UV}{c(t)}$

Table 4.6 Direct Representations of Fluid Circuit Elements (*continued*)

Name	Symbol	SI unit	Frequently used clinical units
Volume	$v(t)$	m^3	L, ml
Pressure	$p(t)$	Pa (=N/m^2)	mm Hg, cm H_2O
Flow rate	$f(t)$	m^3/s	L/min, ml/s
Resistance	R	Pa m^{-3} s	dyn s cm^{-5}, cm H_2O L^{-1} s
Compliance	C	m^3/Pa	ml/mm Hg, ml/cm H_2O
Fluid inductance	I	Pa m^{-3} s^2	—

TABLE 4.7 Basic Variables and Parameters in Fluid Circuits

mathematical description of the resistance reflect Ohm's law for fluid circuits.[3] For this, as well as for the other elements, the direction of the arrow corresponds to a positive flow rate. In the later paragraphs, we will come back to the physics behind this law, and in Sec. 4.7 we will revisit the equivalencies with other electrical circuit elements. The second element contains a nonlinearity, reflecting that below an unstressed volume *UV*, the vessel (or airway) segment is in a collapsed state, without generating pressure. Imagine a balloon that requires a certain volume, before pressure starts building up. Above this volume, the compliance, which is equal to the ratio between change of volume and change of pressure, is assumed to be constant. *UV* is an additional parameter, required to characterize this relationship. In the transmural pressure of the compliance $p_1(t) - p_2(t)$, $p_2(t)$ is often assigned a zero value, representing a reference pressure, for example, ambient pressure. Note that only positive volumes and pressures are taken into consideration for this element. Besides the description of this elastic behavior, the mathematical description of the compliance contains a second equation reflecting conservation of mass: The change of volume is equal to the inflow rate minus the outflow rate. Note that the compliance may also be connected to a single, possibly bidirectional flow rate. The inductor represents inertial phenomena: it takes a pressure difference and a certain amount of time to change flow rate. An open valve has an internal resistance *R*. The valve closes when the pressure across the valve is negative and there is no reverse flow. Two basic active components are also included in Table 4.6. An (ideal) compressor maintains a specified pressure difference regardless of the flow rate through it. Heart chambers can be represented by time-varying compliances (Beneken, 1967; Suga, 1973). The mathematical description of this element is identical to the passive compliance, with the

[3] All depicted relationships are static; also see Table 4.1, but we maintain the notation $v(t)$ for a variable to clearly distinguish it from a parameter *P*.

exception of a time varying $c(t)$. Often, the mathematical description uses the inverse of compliance: elastance:

$$p(t) = e(t)[v(t) - UV] \tag{4.31}$$

where $p(t)$ is the transmural pressure $p_1(t) - p_2(t)$, and with, for example,

$$e(t) = \begin{cases} E_{\min} + (E_{\max} - E_{\min})\sin(\pi t / T_{vs}) & for \quad 0 < t < T_{vs} \\ \\ E_{\min} & (otherwise) \end{cases} \tag{4.32}$$

Figure 4.5 shows $e(t)$, repeated with heart period *HP*. The other parameters of this curve are the duration of ventricular systole, T_{vs}, and the minimum and maximum elastances, E_{min} and E_{max}, respectively.

From a general modeling perspective, we observe that the relationships and mathematical descriptions may introduce assumptions and simplifications in addition to those already included in the conceptual model.

The mathematical relationships for the compliances and for the inductor contain derivatives of volume in and flow rate through the element, respectively. When the elements are included in a network, these are good choices of state variables, and equations of the type of Eq. (4.28) can often be derived. Rules for combining these elements are analogous to the rules for combining electrical elements in circuits, and will be discussed further in Sec. 4.7. There we will also demonstrate how a mathematical model for a complete circuit is derived. In Part II of the book we will use these elements to build conceptual models, such as Fig. 3.10(a), and derive mathematical models from them. Differences between hydraulics and pneumatics

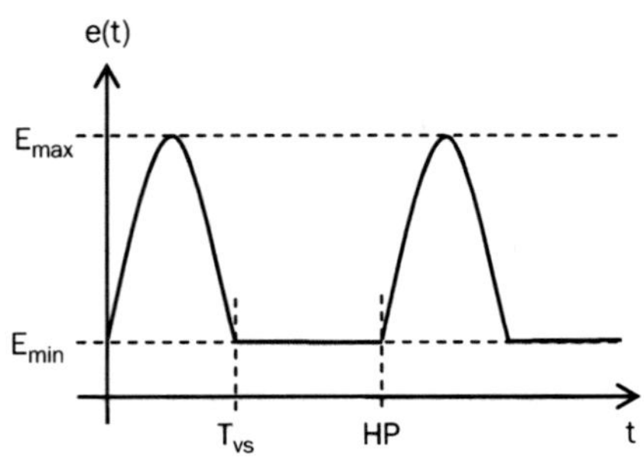

Figure 4.5 Simple cardiac elastance function.

will be discussed there too. We verify that only the mentioned state variables contain information about the past of the system.

A deeper understanding of the circuit elements, depicted relationships, mathematical descriptions, and implicit simplifications can be gained both from further study of physics and "real world" experiential learning. The mathematical description of the resistor, for example, is a form of Poiseuille's law for laminar fluid flow through a uniform straight pipe of radius r and length l

$$f(t) = \pi r^4 \frac{p_1(t) - p_2(t)}{8\eta l} \tag{4.33}$$

where η is the dynamic viscosity. Propelling air or water through tubes of different diameter and length will make you experience this law. Studying Young's modulus and blowing air into more or less compliant balloons will improve understanding of compliance. Newton's second law of motion (see Sec. 4.1) and trying to lift a bucket of water slowly vs. instantaneously ($df(t)/dt \rightarrow \infty$) provides insight into the fluid inductor.

4.5 Direct Representation of Gas Uptake and Distribution

Basic concepts for the description of a gas in a gas mixture or a gas dissolved in a liquid, such as partial pressure, solubility, diffusion, and the ideal gas law, can be found in many introductory texts on chemistry and physics (e.g., Brady, 2000; Haliday, 2001). References to the rather complex binding of the respiratory gases oxygen and carbon dioxide to other molecules in blood will be given in Part II. A direct representation that is fundamental in the modeling of physiological systems involving concentration-type variables is the homogeneous compartment. A number of elementary homogeneous compartments that are frequently used in modeling of gas uptake in the lungs and distribution via the circulation are listed in Table 4.8, together with their mathematical descriptions. The relationships between variables in these compartments are more complex than for fluid circuit elements; therefore, a simple graphical representation cannot be given here. Table 4.9 lists the involved quantities.

Table 4.8 reflects the use of partial pressure, rather than content, as a state variable. This represents expansion of work on inert gases by, for example, Mapleson (1963) and Zwart (1972), rather than a traditional content-based approach to modeling of respiratory gases such as that presented by, for example, Grodins (1967) and Chiari (1997). The advantages of this approach are that it facilitates description of diffusion phenomena and allows for a more unified representation of mixing phenomena in gases and

Description	Symbol	Mathematical description
Ventilated compartment	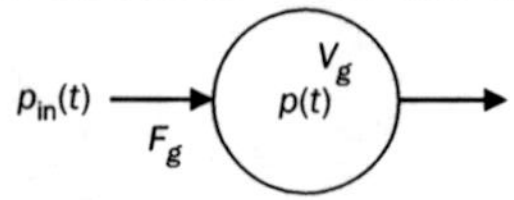	$\frac{dp(t)}{dt} = \frac{F_g}{V_g}[p_{in}(t) - p(t)]$
Perfused compartment, inert gas	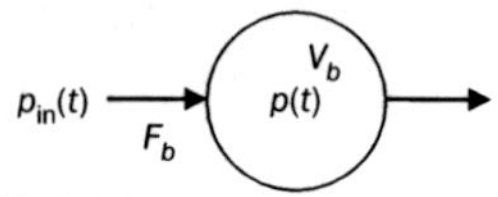	$\frac{dp(t)}{dt} = \frac{F_b}{V_b}[p_{in}(t) - p(t)]$
Perfused compartment, gas with binding	(same symbol, variables, and parameters)	$\frac{dp(t)}{dt} = \frac{F_b\{c_b[p_{in}(t)] - c_b[p(t)]\}}{V_b(dc_b(p)/dp)\vert_{p(t)}}$
Perfused blood-tissue compartment	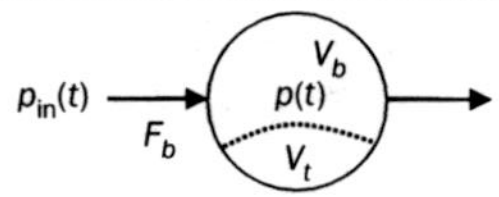	$\frac{dp(t)}{dt} = \frac{F_b\{c_b[p_{in}(t)] - c_b[p(t)]\}}{V_b(dc_b(p)/dp)\vert_{p(t)} + V_t(dc_t(p)/dp)\vert_{p(t)}}$
Perfused blood-tissue compartment with metabolism	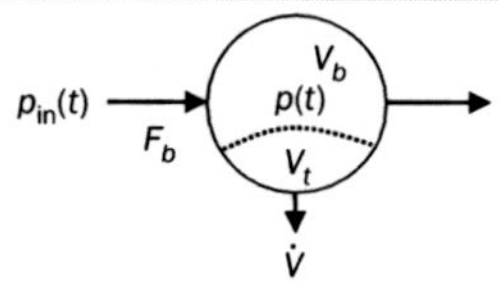	$\frac{dp(t)}{dt} = \frac{F_b\{c_b[p_{in}(t)] - c_b[p(t)]\} - \dot{V}}{V_b(dc_b(p)/dp)\vert_{p(t)} + V_t(dc_t(p)/dp)\vert_{p(t)}}$
Ventilated and perfused gas-blood compartment	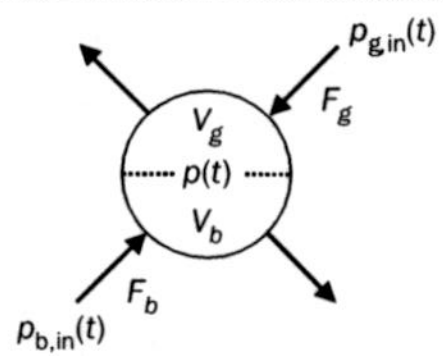	$\frac{dp(t)}{dt} = \frac{F_g(p_{g,in}(t) - p(t)/P_T) + F_b\{c_b[p_{in,b}(t)] - c_b[p(t)]\}}{V_g(1/P_T) + V_b(dc_b(p)/dp)\vert_{p(t)}}$

TABLE 4.8 Direct Representations of Homogeneous Compartments

Name	Symbol	SI unit	Frequently used clinical units
Partial pressure	$p(t)$	Pa (= N m^{-2})	mm Hg
Concentration	c(t)	mol m^{-3}, kg m^{-3}	L or ml of gas at STP or BTPS/L or dl of blood
Gas volume	V_g	m^3	L or ml at STP or BTPS
Blood volume	V_b	m^3	L, ml
Tissue volume	V_t	m^3	L, ml
Gas flow rate	F_g	$m^3\ s^{-1}$	L at STP or BTPS min^{-1}, ml at STP or BTPS s^{-1}
Blood flow rate	F_b	$m^3\ s^{-1}$	L min^{-1}, ml s^{-1}

See main text for STP and BTPS.

TABLE 4.9 Variables, Parameters, and Units in Homogeneous Compartments

fluids, and of inert gases and gases with binding. It brings with it some additional complexity in the form of the derivative of the partial pressure–content relationships. In this chapter we consider that the involved carrier flow rates and volumes are constant and positive. In Chap. 9 we will lift those restrictions—with the exception of positive volumes—and discuss multiple models combining dynamic fluid models with homogeneous compartment models.

The SI units for concentration of a gas in a gas mixture or dissolved in a liquid are mol/m^3 or kg/m^3. In medicine, for a gas in a gas mixture, and by extension for a gas dissolved in a liquid, volume fractions are often used for example, milliliter of oxygen per deciliter of blood. These units are often taken over in physiologic models. They suffer from the following ambiguity: how much gas (in moles or kilogram) a particular volume (milliliter or liter) contains, depends on temperature, pressure, and the amount of water vapor in the gas. As long as the units are carefully specified, this does not constitute a problem. The following conversions from mole to liter apply. For a gas that obeys the ideal gas law:

$$PV = nRT \tag{4.34}$$

with pressure P in Pa, volume V in m^3, the number of moles n, the universal gas constant $R \approx 8.31$ Pa m^3 mol^{-1} K^{-1}, and temperature T in K, 1 mol at the standard temperature $T = 0$ °C ≈ 273 K and standard pressure[4] $P = 100$ kPa occupies approximately

4 Older textbooks use the standard pressure 1 atm = 101.3 kPa.

$$V_1 = \frac{nRT_1}{P} \approx \frac{1 \times 8.31 \times 273}{10^5} \approx 0.0227\,\text{m}^3 = 22.7\,\text{L at STP} \tag{4.35}$$

Applying the ideal gas law, we find that at body temperature T = 37 °C ≈ 310 K, but otherwise identical conditions, the same amount of gas occupies approximately

$$V_2 = \frac{T_2}{T_1} V_1 \approx \frac{310}{273} 22.7 \approx 25.8\,\text{L at BTPD} \tag{4.36}$$

where D stands for dry gas. The vapor pressure of water at body temperature is approximately 47 torr ≈ 0.06 10^5 Pa (Brady, 2001). If the gas is saturated with water vapor, its pressure decreases from the standard pressure to a partial pressure

$$P_{\text{gas}} = P - P_{\text{H}_2\text{O}} \approx (1.00 - 0.06) \times 10^5 = 0.94 \times 10^5\,\text{Pa} \tag{4.37}$$

Again applying the ideal gas law, the same amount of gas, now mixed with water vapor, occupies approximately

$$V_3 = \frac{P}{P_{\text{gas}}} V_2 = \frac{10^5}{0.94 \times 10^5} 25.8 \approx 27.4\,\text{L at BTPS} \tag{4.38}$$

where S stands for saturated with water vapor. For precise models involving inspired gases, ambient temperature, pressure, and relative humidity have to be taken into account.

We will derive the mathematical description of the perfused blood-tissue compartment with metabolism, and then show how it can be simplified to obtain the description of the perfused compartment with inert gas. The amount of gas in the blood-tissue compartment consists of gas dissolved in blood and gas dissolved in tissue. Both forms are at the same partial pressure, but have different and potentially nonlinear binding characteristics:

$$a(t) = V_b c_b\left[p(t)\right] + V_t c_t\left[p(t)\right] \tag{4.39}$$

Mass transport is equal to the product of carrier (blood) flow rate and solute concentration. For example, in the compartment inflow:

$$F_b c_b\left[p_{in}(t)\right] \tag{4.40}$$

The change of amount depends on inflow, outflow, and metabolism, formalized in the mass balance:

$$\frac{da(t)}{dt} = F_b c_b\left[p_{in}(t)\right] - F_b c_b\left[p(t)\right] - \dot{V} \tag{4.41}$$

To write this mass balance in terms of partial pressures, we first take the derivative of Eq. (4.39):

$$\begin{aligned}\frac{da(t)}{dt} &= V_b \left.\frac{dc_b(p)}{dp}\right|_{p(t)} \frac{dp(t)}{dt} + V_t \left.\frac{dc_t(p)}{dp}\right|_{p(t)} \frac{dp(t)}{dt} \\ &= \left[V_b \left.\frac{dc_b(p)}{dp}\right|_{p(t)} + V_t \left.\frac{dc_t(p)}{dp}\right|_{p(t)} \right] \frac{dp(t)}{dt}\end{aligned} \tag{4.42}$$

Combining Eqs. (4.41) and (4.42) and reorganizing terms, we find the mathematical description included in Table 4.8:

$$\frac{dp(t)}{dt} = \frac{F_b\left\{c_b\left[p_{in}(t)\right] - c_b\left[p(t)\right]\right\} - \dot{V}}{\left[V_b \left.\frac{dc_b(p)}{dp}\right|_{p(t)} + V_t \left.\frac{dc_t(p)}{dp}\right|_{p(t)} \right]} \tag{4.43}$$

For the perfused compartment with inert gas, tissue volume and metabolism are zero, and content is found via Henry's law, for example, in blood:

$$c_b\left[p(t)\right] = \alpha_b p(t) \tag{4.44}$$

where α_b is the solubility coefficient for the considered gas in blood.[5] For inert gases, the derivative of the partial pressure–content relationship is simply:

$$\frac{dc_b(p)}{dp} = \alpha_b \tag{4.45}$$

Mass transport simplifies to:

$$F_b \alpha_b p_{in}(t) \tag{4.46}$$

[5] Here we follow the standard notation of the parameter α in lower case.

The solubility coefficients cancel out and Eq. (4.43) can be simplified to the result listed in Table 4.8:

$$\frac{dp(t)}{dt} = \frac{F_b}{V_b}\left[p_{in}(t) - p(t)\right] \tag{4.47}$$

For the derivation of mathematical descriptions of ventilated compartments, we need to take into account that the amount of gas in a gas mixture is equal to:

$$a(t) = V_g \frac{p(t)}{P_T} \tag{4.48}$$

where P_T represents the total pressure. The mass transport of gas in a gas mixture is equal to, for example:

$$F_g \frac{p_{in}(t)}{P_T} \tag{4.49}$$

In the description of the ventilated compartment, Table 4.8, the parameter P_T cancels out, but we do find it back in the description of the ventilated and perfused compartment, which is a simple lung model. Binding characteristics of specific respiratory gases will be presented in Chap. 9. Conceptual models of concentration-type models often include several connected homogeneous compartments [see, for example, Fig. 3.10(b)]. Table 4.10 lists several of those connections and their mathematical descriptions.

Concatenation and multiple outflows have straightforward descriptions. The mathematical description of the volumeless confluent reflects conservation of mass of the carrier fluid, instantaneous mixing and conservation of solute mass (the equation reflects blood carrying gas with binding), and computation of the inverse of the partial pressure–content relationship. The latter is only computed if partial pressure is explicitly required. The derivation of the mathematical description of a compartment with multiple (blood) inflows is very similar to the derivation for the perfused blood-tissue compartment with metabolism. It is equivalent to a confluent followed by a compartment. In Part II of the book we will use these elements and connections to build conceptual models [such as Fig. 3.10(b)] and derive mathematical models from them. Models for important respiratory phenomena such as dead space and shunt will be introduced there.

Description	Symbol	Mathematical description
Confluent	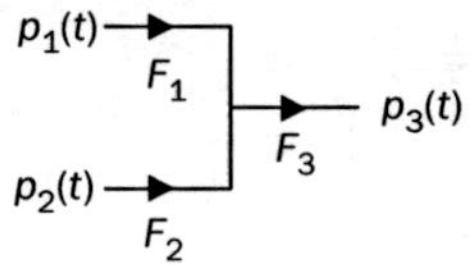	$F_3 = F_1 + F_2$ $c_3(t) = \dfrac{F_1 c_b[p_1(t)] + F_2 c_b[p_2(t)]}{F_1 + F_2}$ $p_3(t) = c_b^{-1}[c_3(t)]$
Concatenation	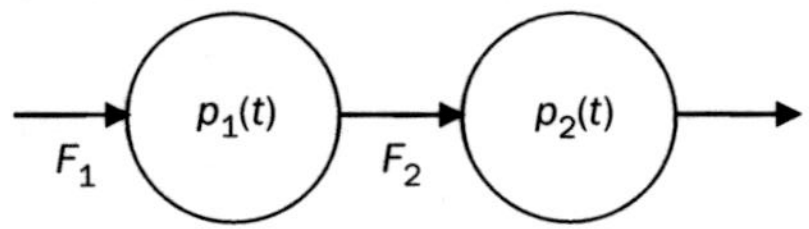	$F_2 = F_1$ $p_{in,2}(t) = p_1(t)$
Multiple inflows	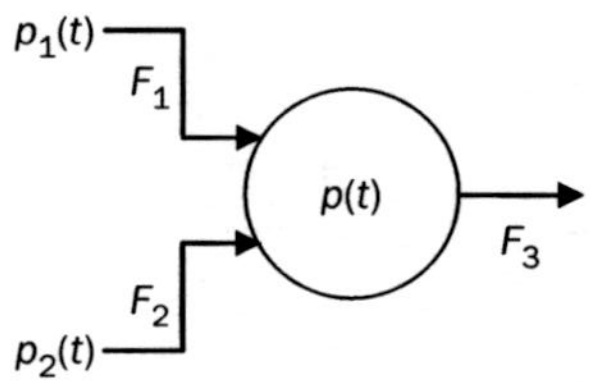	$F_3 = F_1 + F_2$ $\dfrac{dp(t)}{dt} = \dfrac{F_1\{c_b[p_1(t)] - c_b[p(t)]\} + F_2\{c_b[p_2(t)] - c_b[p(t)]\}}{V_b(dc_b(p)/dp)\big\vert_{p(t)}}$
Multiple outflows	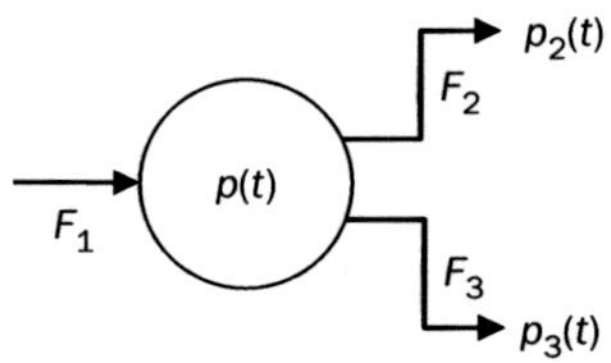	$F_2 + F_3 = F_1$ $p_3(t) = p_2(t) = p(t)$

Table 4.10 Connecting Homogeneous Compartments

4.6 Direct Representation of Simple Transfers in the Nervous System

The central and peripheral nervous systems process many physiological variables. For example, the firing frequency of afferent neurons from peripheral chemoreceptors to the brain stem $f_a(t)$ depends on the partial pressure of oxygen in arterial blood $p_aO_2(t)$. The brain stem processes this signal, resulting in the firing frequency of efferent neurons $f_e(t)$ influencing, for example, heart rate hr(t) (Fig. 4.6).

The processing going on in each individual block can often be described by one or more simple mathematical operations. A number of elementary operations that are frequently used in modeling of transfers in the nervous systems are shown in Table 4.11, together with their graphical representation. The involved quantities can be quite diverse, and are therefore represented by generic input $u(t)$ and output $y(t)$ variables and generic parameters, such as an offset Y_0 and time constant τ.[6]

The first five elementary operations are static, and their mathematical representation is in the form of Eq. (4.29). The last operation is dynamic and its mathematical representation is in the form of Eq. (4.28). The output equation for this element would simply be $y(t) = x(t)$. Although special symbols exist for some simple functions, such as a sum or integrator element, we chose here to represent all these relationships by a simple block with input and output variables. The relationship or the mathematical description can be included in the block [see Fig. 3.10(c)]. The signal processing and control engineering literature provide more elaborate system transfers (Roberts, 2003). Connecting rules are simple and identical to those of general block diagrams. An input has to have a well-defined, single origin. Outputs can be sent to one or more other blocks, resulting in series or parallel combinations. The sum block combines two or more inputs. In Part II of the book we will use these elements and connecting rules to build conceptual models of systems (such as in Fig. 4.6).

4.7 Electrical Analogs and State Variable Models of Circuits

For a representation to be qualified as an analog of a given system, the underlying mathematical models should be the same. In that sense, and after demonstrating that the two systems have identical underlying mathematical models, Fig. 4.1 is a mechanical

[6] Here we follow the standard notation of the parameter τ in lower case.

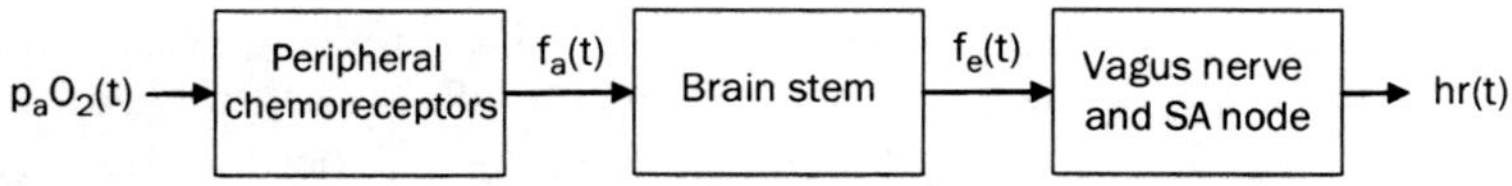

FIGURE 4.6 Block diagram of processes involved in chemoreceptor control of heart rate. See main text for symbols.

Name	Relationship	Mathematical description
Sum	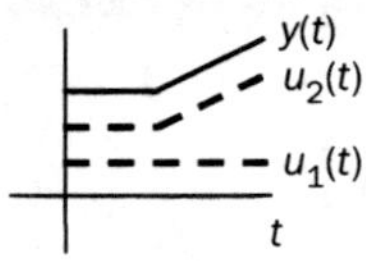	$y(t) = u_1(t) + u_2(t)$
Gain	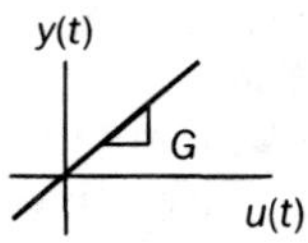	$y(t) = Gu(t)$
Gain with offset	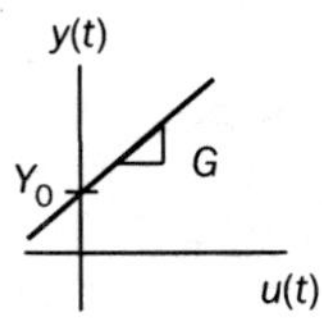	$y(t) = Gu(t) + Y_0$
Piecewise linear function	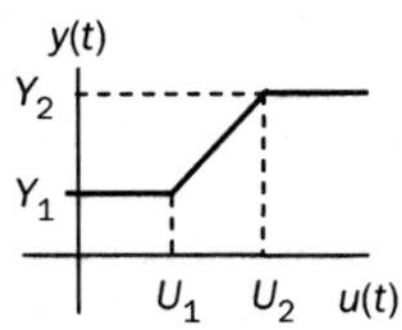	if $u(t) \le U_1$ $y(t) = Y_1$ else if $u(t) < U_2$ $y(t) = Y_1 + \frac{Y_2 - Y_1}{U_2 - U_1}(u(t) - U_1)$ else $y(t) = Y_2$
Sigmoid curve	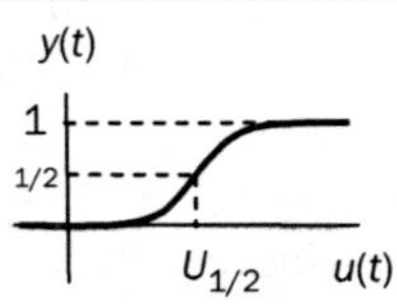	$y(t) = \frac{u(t)^\gamma}{u(t)^\gamma + U_{1/2}^\gamma}$
First order transfer	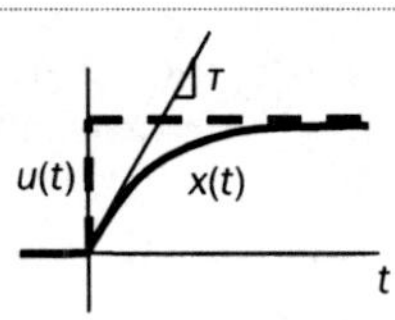	$\frac{dx(t)}{dt} = \frac{u(t) - x(t)}{\tau}$

TABLE 4.11 Representations of Simple System Transfers

analog for the electrical system of Fig. 4.3, and Fig. 4.3 is an electrical analog for the mechanical system of Fig. 4.1. In the past, electrical analogs were often used to represent hemodynamics and ventilatory mechanics, and even gas exchange, but we recommend the use of direct representations to establish conceptual models. This greatly facilitates the dialog between life scientists and modelers, and avoids a number of problems related to equivalencies between variables and parameters and associated unit conversions. Analogs nevertheless still have useful applications in physiologic modeling. In the first place, they may provide additional insights in a given system. Consider, for example, the equivalencies between the spring in Fig. 4.1 (Eq. 4.1), the capacitor in Fig. 4.3 (Eq. 4.7), and the compliance in Table 4.5. What we learn about one system is extended to an understanding of the other system. However, when making such extensions, we need to be aware of the simplifications underlying idealized components. For example, there is no equivalent phenomenon to the mentioned breakaway forces for the frictional element in Fig. 4.1 for current flow through a resistor. A somewhat related application of an analog is that it may be in a domain that the modeler is more familiar with. A mechanical, or mechanically oriented, engineer may prefer to work with Fig. 4.1 as a representation of the electrical system, once it is established as an analog. Similarly, a transformation into another domain may provide a convenient and more intuitive visualization of a system, or of its mathematical model, that would otherwise remain difficult to understand. For example, Nikkelen et al. (1998) used a hydraulic analog of the dynamic, nonlinear processes of drug distribution and drug effect on muscle relaxation. An accelerated time simulation of this analog was used in teaching pharmacokinetics and pharmacodynamics to anesthesia residents. A special form of direct representation, sometimes also called an analog, is a linear equivalent circuit. For example, an electrical equivalent circuit for the electrode–skin interface may provide useful insights in the main dynamics of this system. A final application of analogs is based on the fact that, in some physical domains, certain techniques are worked out further than in others. Here we will present network reduction techniques for parallel and in series electrical circuit elements, and a more extensive electrical circuit example. Via analogy, the same concepts also apply to direct representations of hydraulic and pneumatic circuits. In preparation of that, and to facilitate access to older biomedical modeling literature, Tables 4.12 and 4.13 present electrical circuit elements and quantities in the order of equivalencies to Tables 4.6 and 4.7.

Name	Symbol	Relationship	Mathematical description
Resistor	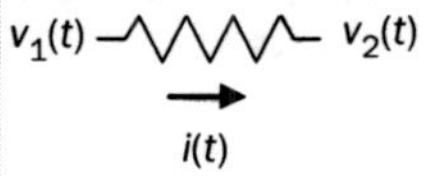	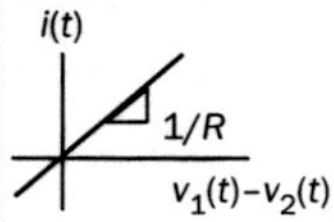	$i(t)=\dfrac{v_1(t)-v_2(t)}{R}$
Capacitor	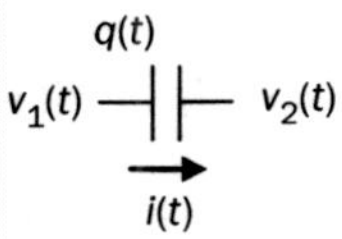	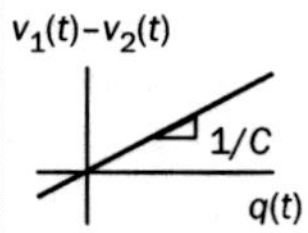	$v_1(t)-v_2(t)=\dfrac{q(t)}{C}$ $i(t)=\dfrac{dq(t)}{dt}=C\dfrac{d(v_1(t)-v_2(t))}{dt}$
Inductor	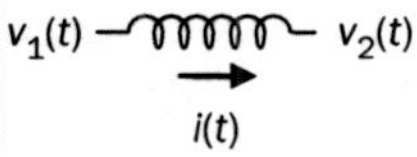	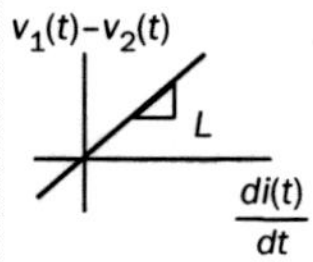	$v_1(t)-v_2(t)=L\dfrac{di(t)}{dt}$
Diode (in series with a resistor)	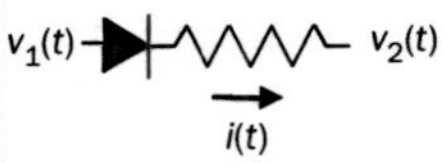	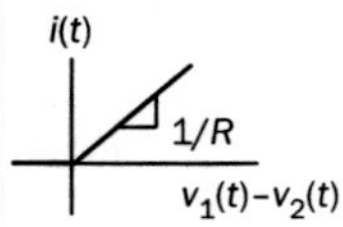	if $v_1(t)-v_2(t)>0$ $i(t)=\dfrac{v_1(t)-v_2(t)}{R}$ else $i(t)=0$
Voltage source	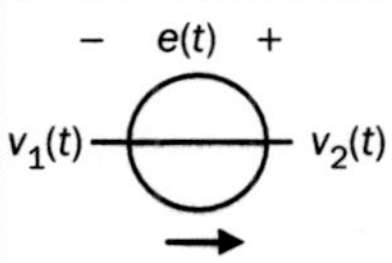	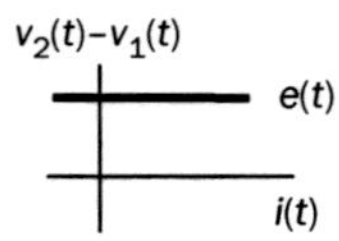	$v_2(t)=v_1(t)+e(t)$

TABLE 4.12 Electrical Analogs to the First Five Fluid Circuit Elements of Table 4.6

Name	Symbol	SI unit
Charge	$q(t)$	C (=A·s)
Voltage	$v(t)$	V (=N m A^{-1} s^{-1})
Current	$i(t)$	A
Resistance	R	Ω (= V A^{-1})
Capacitance	C	F (= A·s V^{-1})
Inductance	L	H (= V·s A^{-1})

TABLE 4.13 Basic Variables and Parameters in Electrical Circuits

Unlike the equivalent compliance, the capacitor has a linear model, and allows for negative charges. The most commonly used state variable is voltage, equivalent to pressure, rather than charge, equivalent to volume.

In modeling physiological and other systems we frequently encounter in series and parallel circuit elements, for example, in series upper airway and bronchial resistances to gas flow, or parallel left and right lung compliances. If volumes or flow rates (charges or currents) on or through the individual elements are not of interest, then these elements can be combined, simplifying the model without loss of generality. Table 4.14 shows several rules for combining linear electrical resistors and capacitors.

These equivalencies can be derived using component descriptions and the Kirchhoff laws. Kirchhoff's voltage law (KVL) was introduced in Sec. 4.1. Kirchhoff's current law (KCL) specifies that the sum of the currents flowing into a node is equal to the sum of currents flowing out of a node. For example, assuming initially uncharged capacitors in series, if a charge $q(t)$ is injected in capacitor C_1, then by KCL, the same current and charge enter C_2. Using KVL and the component description, Table 4.12, we find the voltage across the two capacitors

$$v_1(t) - v_2(t) = \frac{q(t)}{C_1} + \frac{q(t)}{C_2} \tag{4.50}$$

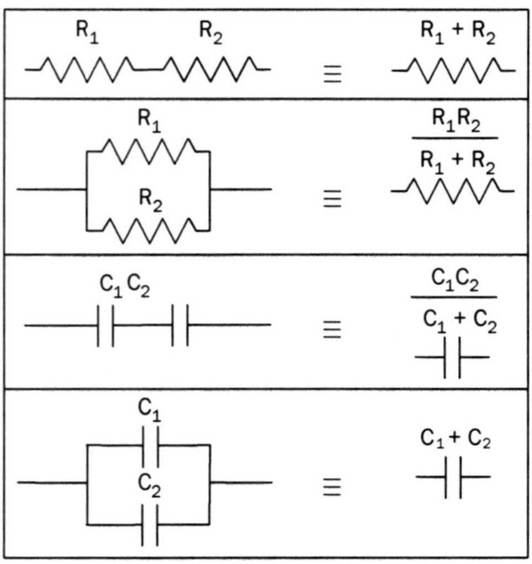

TABLE 4.14 In Series and Parallel Resistors and Capacitors

or

$$\frac{v_1(t) - v_2(t)}{q(t)} = \frac{1}{C_1} + \frac{1}{C_2} = \frac{C_1 + C_2}{C_1 C_2} \tag{4.51}$$

and the total capacitance is

$$C = \frac{q(t)}{v_1(t) - v_2(t)} = \frac{C_1 C_2}{C_1 + C_2} \tag{4.52}$$

Figure 4.7 shows a fluid circuit, part of an elementary vascular model, and its electrical analog.

We will derive a state variable model for this network in the domain of the analog. Remember that the goal of a state variable model is to express the change of the state variables, which we will choose to be the charges on the capacitors, as a function of the same state variables, and of the input voltage $e(t)$. At a specific time t, the voltages across the capacitors can be expressed as a function of their charges, using the component description in Table 4.12.

$$v_1(t) = \frac{q_1(t)}{C_1} \tag{4.53}$$

$$v_2(t) = \frac{q_2(t)}{C_2} \tag{4.54}$$

$$v_3(t) = \frac{q_3(t)}{C_3} \tag{4.55}$$

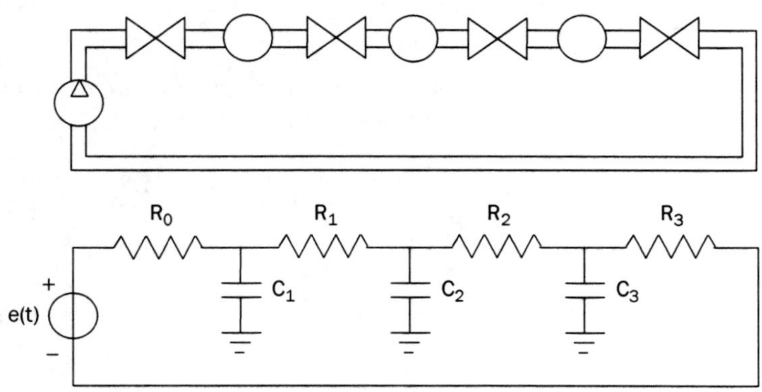

FIGURE 4.7 Fluid circuit and its electrical analog.

The currents through the resistors R_1 and R_2 depend on the voltages according to Ohm's law, also included in Table 4.12.

$$i_1(t) = \frac{v_1(t) - v_2(t)}{R_1} \tag{4.56}$$

$$i_2(t) = \frac{v_2(t) - v_3(t)}{R_2} \tag{4.57}$$

We find an expression for the current $i_3(t) = i_0(t)$ through the resistors R_3 and R_0 via the KVL for the closed circuit: ground, C_3, R_3, source, R_0, C_1, and back to ground, taking into account the directed voltages.

$$v_3(t) - R_3 i_3(t) + e(t) - R_0 i_3(t) - v_1(t) = 0 \tag{4.58}$$

or

$$i_3(t) = i_0(t) = \frac{v_3(t) + e(t) - v_1(t)}{R_0 + R_3} \tag{4.59}$$

The change in charge on each capacitor is equal to the difference between incoming and outgoing currents.

$$\frac{dq_1(t)}{dt} = i_0(t) - i_1(t) \tag{4.60}$$

$$\frac{dq_2(t)}{dt} = i_1(t) - i_2(t) \tag{4.61}$$

$$\frac{dq_3(t)}{dt} = i_2(t) - i_3(t) \tag{4.62}$$

By substituting the voltage equations in the current equations, and the current equations in the charge equations, we find:

$$\frac{dq_1(t)}{dt} = \frac{q_3(t)/C_3 + e(t) - q_1(t)/C_1}{R_0 + R_3} - \frac{q_1(t)/C_1 - q_2(t)/C_2}{R_1} \tag{4.63}$$

$$\frac{dq_2(t)}{dt} = \frac{q_1(t)/C_1 - q_2(t)/C_2}{R_1} - \frac{q_2(t)/C_2 - q_3(t)/C_3}{R_2} \tag{4.64}$$

$$\frac{dq_3(t)}{dt} = \frac{q_2(t)/C_2 - q_3(t)/C_3}{R_2} - \frac{q_3(t)/C_3 + e(t) - q_1(t)/C_1}{R_0 + R_3} \tag{4.65}$$

or

$$\begin{bmatrix} \frac{dq_1(t)}{dt} \\ \frac{dq_2(t)}{dt} \\ \frac{dq_3(t)}{dt} \end{bmatrix} =$$

$$\begin{bmatrix} -\frac{1}{C_1}\left(\frac{1}{R_0+R_3}+\frac{1}{R_1}\right) & \frac{1}{C_2R_1} & \frac{1}{C_3}\frac{1}{R_0+R_3} \\ \frac{1}{C_1R_1} & -\frac{1}{C_2}\left(\frac{1}{R_1}+\frac{1}{R_2}\right) & \frac{1}{C_3R_2} \\ \frac{1}{C_1}\frac{1}{R_0+R_3} & \frac{1}{C_2R_2} & -\frac{1}{C_3}\left(\frac{1}{R_2}+\frac{1}{R_0+R_3}\right) \end{bmatrix} \begin{bmatrix} q_1(t) \\ q_2(t) \\ q_3(t) \end{bmatrix} + \begin{bmatrix} \frac{1}{R_0+R_3} \\ 0 \\ \frac{-1}{R_0+R_3} \end{bmatrix} e(t) \tag{4.66}$$

which is indeed a state equation of the form of Eq. (4.26)

$$\dot{x}(t) = Ax(t) + Bu(t) \tag{4.67}$$

Using the equivalencies between electrical and hydraulic quantities as listed in Tables 4.13 and 4.7, respectively, and substituting the external voltage by an externally applied pressure, we can transform this model into a model for the analog hydraulic circuit. For this relatively simple circuit, we could also have derived the mathematical model directly from the fluid circuit and its component equations. If we choose to include nonlinear compliances, the resulting model will be of the form of Eq. (4.28). Note that the state variables in this example are dependent: the sum of the charges on the capacitors or of the volumes in the compliances is constant. The model could therefore be simplified. However, if a leakage current or blood loss is included, this would no longer be the case, and the same systematic model derivation applies.

4.8 General Observations on Mathematical Models and Parameter Estimation

The mathematical model has obvious advantages for system analysis, and by being unambiguous, concise (yet complete), and self-consistent (Khoo, 1999) it is the best form to archive knowledge about the system. A mathematical model in state variable notation is fully characterized by input and output variables, state and output equations, and numerical values of parameters and initial state variables. A code implementation is necessary to obtain simulation results, but does not have the properties of a mathematical model that make it convenient for analysis and archiving knowledge. We do not advocate the use of techniques, however convenient, that skip the mathematical model and code implementation steps, and go directly from "drag-and-drop" diagrams of system structure and function to simulation results; the overall block diagram, conceptual model, mathematical model, and simulation results provide complementary perspectives on system behavior, which are most insightful when established separately and then considered in conjunction.

For white-box models, the parameters are known, and involve physical constants or measurable quantities, for example, the universal gas constant or total blood volume. For gray-box models, some of the parameters may have to be estimated. The methods to do so can be divided into two broad categories. One approach is to search the scientific literature for numerical values of model parameters for specific structures or processes. These parameters may have to be scaled from different human or even animal populations. See Goodwin et al. (2004) for an example of this process applied to a cardiovascular model. An alternative method is to treat a model with a few remaining unknown parameters as a black-box model, and apply standard parameter estimation techniques (Ljung, 1999). If only a single model parameter remains to be estimated, and if resulting limited accuracy is acceptable, a point estimate based on a single observable response may be sufficient. See Van Meurs et al. (1998) for an example of this process applied to a pharmacological model.

A possible concern with state variable models is that the set of independent state variables may not be unique. For example, we could have used the voltages across the capacitors in Fig. 4.7 instead of their charges as state variables. Therefore, there may be several valid state variable models for the same system. A second concern has to do with the time-domain character of state variable models. There is no equivalent to the simple multiplication of transfer functions for calculation of the response of two systems in series. This is a concern for system analysis and design, but not for simulation. For simulation purposes, two state variable models can easily be connected by specifying:

$$u_2(t) = y_1(t) \tag{4.68}$$

where $y_1(t)$ is the output of the first system, and $u_2(t)$ is the input of the second system. This puts some constraints on the order in which equations have to be computed. We will investigate this topic further in the next chapter.

In Secs. 4.1 and 4.2 we showed in a simple example how the same system can be represented by an ordinary differential equation and a state variable model, respectively. For a more formal treatment of the relationships between these mathematical models, and the relationships with models in the form of the impulse response, frequency response function, and transfer function, we again point to Ljung (1999).

Review Problems

4.1 Design a conceptual model for the uncontrolled cardiovascular subsystem of Review Problem 3.1 in the form of a fluid circuit diagram. Take into account the requirements formulated as part of Review Problem 2.1. Derive a mathematical model assuming linear fluid circuit elements and a simple time-varying elastance, as introduced in Sec. 4.4. Search the scientific literature for numerical values of model parameters for the selected patient population. See Goodwin et al. (2004) for an example of this process.

4.2 Derive a state variable model for the translational mechanical system of Fig. 4.1 with the state variables extension of the spring $x(t)$, and velocity of the mass $v(t)$.

4.3 The figure below shows a left ventricular pressure-volume curve, simulated with an elastance function similar to the one of Eq. (4.31) and Fig. 4.5 (Goodwin, 2004).

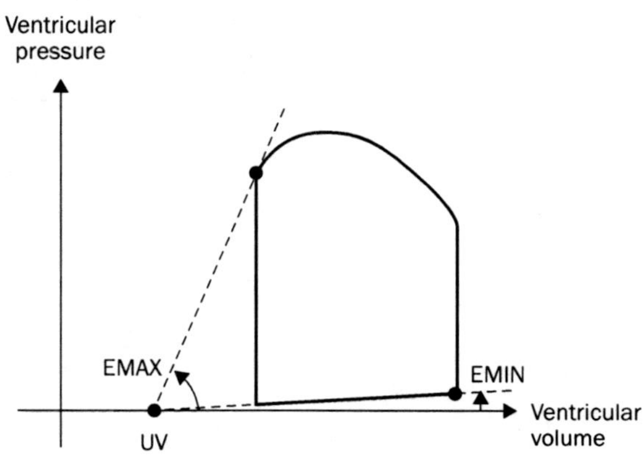

FIGURE 4.8 Graph for review problem 4.3.

Identify the phases of the cardiac cycle in this diagram: filling, isovolumetric contraction, ejection, and isovolumetric relaxation.

References and Further Reading

Athans M. *Systems, networks and computations: Multivariable methods.* New York: McGraw-Hill, 1974.

Beneken JEW, DeWit B. A physical approach to hemodynamic aspects of the human cardiovascular system. In: Reeve EB, Guyton AC, eds. *Physical bases of circulatory transport: regulation and exchange.* Philadelphia: Saunders; 1967:1–45.

Brady JE, Russell JW, Holum JR. *Chemistry: Matter and its changes.* 3rd ed. New York: John Wiley & Sons; 2001.

Chiari L, Avanzolini G, Ursino M. A comprehensive simulator of the human respiratory system: validation with experimental and simulated data. *Ann Biomed Eng.* Nov-Dec 1997; 25(6):985-99.

Cutnell JD, Johnson KW. *Physics*, 4th ed. New York: John Wiley & Sons; 1998.

Goodwin JA, van Meurs WL, Sá Couto CD, Beneken JEW, Graves SA. A model for educational simulation of infant cardiovascular physiology. *Anest Analg.* Dec 2004; 99(6):1655-64.

Grodins FS, Buell J, Bart AJ. Mathematical analysis and digital simulation of the respiratory control system. *J Appl Physiol.* Feb 1967; 22(2):260-76.

Halliday D, Resnick R, Walker J. *Fundamentals of physics.* 6th ed. New York: John Wiley & Sons; 2001.

Khoo MCK. *Physiologic control systems, analysis, simulation, and estimation.* Piscataway: Wiley-IEEE Press; 1999.

Ljung L. *System identification, theory for the user.* 2nd ed. Upper Saddle River, NJ: Prentice Hall; 1999.

Mapleson WW. An electric analogue for uptake and exchange of inert gases and other agents. *Appl Physiol.* Jan 1963; 18:197-204.

Nikkelen E, van Meurs WL, Öhrn MAK. Hydraulic analog for simultaneous representation of pharmacokinetics and pharmacodynamics: Application to vecuronium. *J Clin Monit Comput.* Jul 1998; 14 (5):329-37.

Roberts MJ. *Signals and systems: Analysis of signals through linear systems.* Boston: McGraw Hill; 2003.

Suga H, Sagawa K, Shoukas AA. Load independence of the instantaneous pressure-volume ratio of the canine left ventricle and effects of epinephrine and heart rate on the ratio. *Circ Res.* 1973; 2:314 –22.

Van Meurs WL, Nikkelen E, Good ML. Pharmacokinetic-pharmacodynamic model for educational simulations. *IEEE Trans Biomed Eng.* May 1998; 45(5):582-90.

Zwart A, Smith NT, Beneken JE. Multiple model approach to uptake and distribution of halothane: the use of an analog computer. *Comput Biomed Res.* Jun 1972; 5(3):228-38.

CHAPTER 5
Software Implementation

In this chapter we present a simple discrete-time approximation of the continuous-time state equation, based on the definition of the derivative. Algorithms for off-line and (pseudo) real time software implementation of a state variable model are given. Grouping and sequencing of model equations is of particular importance when connecting state variable models. To avoid circular references when feedback occurs, at least one of the models should not have immediate throughput from the input to the output. Different integration step sizes in two connected models can be accommodated via separate branches in an algorithm.

Input–output requirements are formulated in the form of an overall model block diagram, and an explicit list of input and output variables (see Chap. 2). Internal system structure and functions to be included in the model are reflected in a conceptual model, which can take the form of a block diagram of subsystems and/or of component diagrams (Fig. 3.10). For the subsystems, a mathematical model in the form of a continuous-time state variable formulation can often be formulated (Chap. 4).

$$\dot{x}(t) = f(x(t), u(t), t) \tag{5.1}$$

$$y(t) = g(x(t), u(t), t) \tag{5.2}$$

In this chapter we will see how these equations can be implemented in software on a digital (discrete-time) computer, and how subsystems can be combined. We will only make programming language–independent observations.

5.1 Discretization of the Continuous-Time State Equation

The output equation (5.2) can be implemented "as is." A simple discrete-time code implementation of the state equation (5.1) is based on the definition of the derivative:

$$\dot{x}(t) = \lim_{T \to 0} \frac{x(t+T) - x(t)}{T} \tag{5.3}$$

Combining Eqs. (5.1) and (5.3) we find

$$x(t+T) \approx x(t) + T f(x(t), u(t), t) \tag{5.4}$$

In its definite form Eq. (5.4) can be initiated with $x(t_0)$ and applied iteratively for $t = t_0, t_0 + T, t_0 + 2T, \ldots$ or $t = t_0 + nT$, where n is the *discrete time* and T the *integration step size*. T should be small as compared to the fastest system dynamics to lead to a good approximation of the derivative and to precise integration, but not so small that the simulation time becomes prohibitively long or that the accuracy of the increments of the state or time variables is limited by word length. This basic method of integration is called the *Euler forward method*. Textbooks on numerical integration discuss this method, as well as more accurate and efficient methods (see, for example, Chapra, 2010). For certain models, if the input $u(t)$ is constant over the time interval T, an exact prediction can be computed (see Ljung, 1999).

Throughout this chapter we will assume that the *sampling interval* of $u(t)$ is identical to the integration step size. Reconstruction of $y(t)$ from its samples, again assumed to be spaced at intervals that are identical to the integration step size, will involve interpolation. Interpolation routines, for example, zero-order hold or linear interpolation, are often implicit to hardware implementations or plotting routines, and will greatly affect the frequency content and appearance of a signal.

5.2 Basic Algorithms for Implementation of the Discrete-Time State Variable Model

We first consider a simple algorithm for off-line or non–real time simulation of a state variable model. At the start of the program, a number of quantities need to be set or initialized: the integration step size T, the simulation start and stop times $t = t_0$ and t_f, the model parameters, and the initial state $x(t_0)$ (Fig. 5.1).

We assume that input $u(t)$ is available for $t \geq t_0$, either as an array of values, or in the form of an analytical function. Then the output equation (5.2) can be computed for time $t = t_0$. Note the use of ":=" to indicate an assignment. It is possible that $t_f = t_0$, in which case the simulation is over, and simulation results can be plotted. In this case the "plot" would be reduced to a single value for all variables of interest. An alternative to plotting data inside the program is to save them, and plot them later with another program. If the simulation time is not over, the state prediction

$$x(t+T) := x(t) + T f(x(t), u(t), t) \tag{5.5}$$

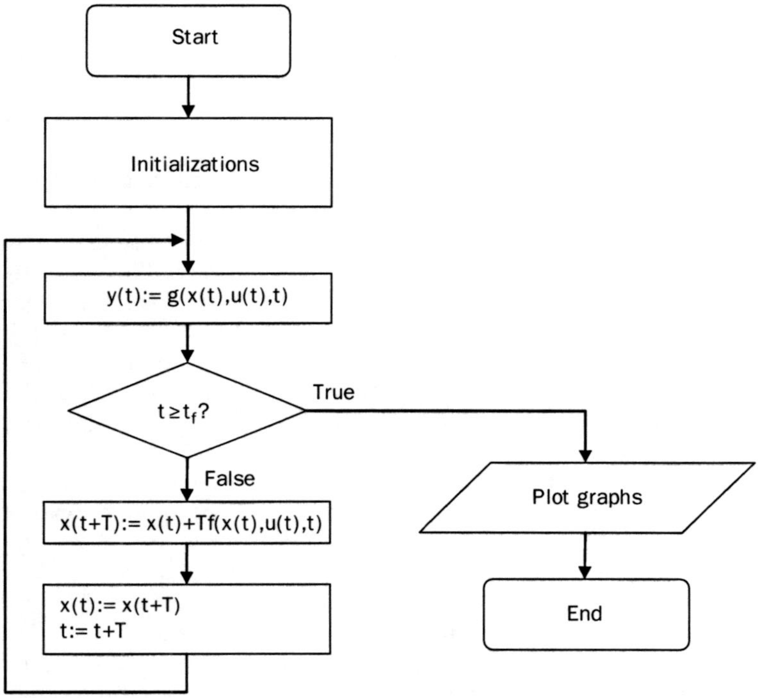

FIGURE 5.1 Flow diagram for off-line software implementation of a state variable model.

can be computed. Here, the assignment takes its full meaning: For a vector of state variables, $x(t + T)$ is a different (code) variable from $x(t)$. Otherwise, for example, for a second-order system, the computation of $x_1(t + T)$ would be based on $x_1(t)$ and $x_2(t)$, and the computation of $x_2(t + T)$ on $x_1(t + T)$ and $x_2(t)$. In preparation for the next iteration, the state variable and time need to be updated:

$$x(t) := x(t+T) \tag{5.6}$$

$$t := t+T \tag{5.7}$$

In this simple algorithm there is no specific assignment for discrete time n. However, in practical code implementations, branching and storage and retrieval of data may be facilitated by maintaining n. We will continue to refer to continuous time t in the assignments, but note that Eqs. (5.2), (5.5), (5.6), and (5.7) are often written as follows:

$$y(n) := g(x(n), u(n), n) \tag{5.8}$$

$$x(n+1) := x(n) + T f(x(n), u(n), n) \quad (5.9)$$

$$x(n) := x(n+1) \quad (5.10)$$

$$n := n+1 \quad (5.11)$$

Unlike what is suggested by this notation, in the algebraic functions t is substituted by nT, not by n.

With the above algorithm, the simulation will run as fast as the model, the integration step size, the structure of the code, the programming language, and the computer hardware allow for. However, several simulation applications require a model to run in real time, or in a specified acceleration or deceleration of real time. For demonstrations, pseudo–real time, a reasonable approximation of real time on a specific computer, may be sufficient. The algorithm of Fig. 5.1 can be expanded as indicated in Fig. 5.2 to meet such requirements. After initializations, the computer "waits" for continuous time to be an integer multiple of the integration step size. This can be implemented via interrupts or scheduling algorithms, or in pseudo–real time via a computer specific cycle in the *False* branch that takes a time $T - T_{ac}$ to run, where T_{ac} is the average computation time for the remainder of the loop. Real time input and output should be grouped around the output equation, and the ensemble should occur immediately after the timing routine to minimize potentially variable time lag with respect to time t. The end of the simulation may now depend on specific user commands. Based on simulation requirements, these algorithms may be expanded with file, user interface, or peripheral device input and output.

5.3 Model Code Verification

The above algorithms, when correctly implemented in software, should avoid problems due to incorrect sequencing of model equations. But when writing software to implement a mathematical model, several other errors can occur: The equations may not be implemented correctly, the numerical values for parameters or the initial state may be incorrect, or the integration step size may be inadequate. Basic precautions include matching names of quantities between the mathematical model and the software, and careful verification by a qualified, preferably independent, person. Some differences in symbols are to be expected, as the mathematical model uses compact, standard mathematical and physiological notation, whereas longer names may have to be used in programming languages, which usually do not have the possibility to use subscripts or superscripts. Remaining errors may be detected when verifying model code functionality via selected simulation results. For specific

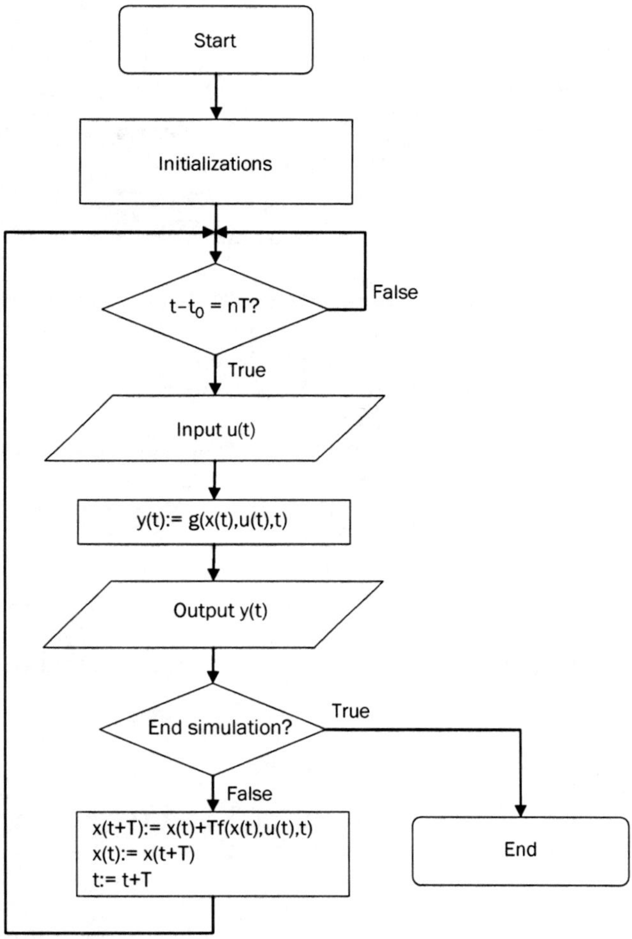

FIGURE 5.2 Flow diagram for (pseudo) real-time software implementation of a state variable model.

inputs, steady state or dynamic responses may have analytical solutions. Simulated numerical values or curves can be compared to those. If the model code is part of a larger, more complex software project, an effort should be made to keep it modular, so that it can be tested independently. In the next chapter we will come back to model code verification as part of the more encompassing model verification and validation.

5.4 Connecting State Variable Models

Now assume two models that interact, with part of the input to the second model coming from the first (Fig. 5.3).

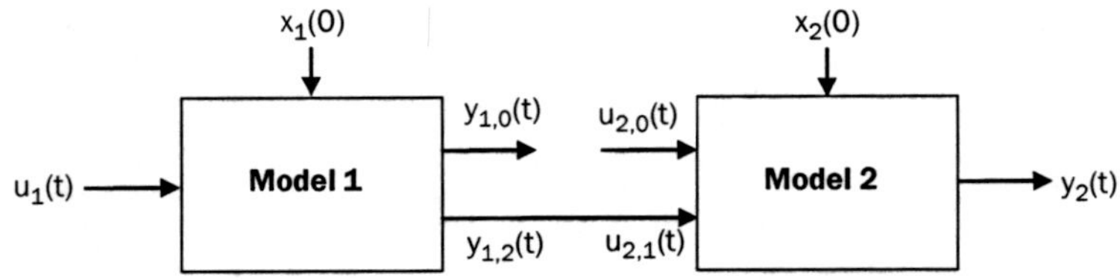

FIGURE 5.3 Block diagram of two connected models.

For example, consider lung mechanics driving pulmonary gas exchange (Fig. 3.5). To simplify the discussion of grouping and sequencing model equations, we will assume that all output from Model 1 goes to Model 2, and that Model 2 does not have other inputs, leading to the block diagram (of Fig. 5.4).

Let us further assume that for the two systems we want to maintain separate mathematical models and code in distinct segments.[1] Initially, we assume that the integration step sizes for the two models are identical:

$$T_1 = T_2 = T \tag{5.12}$$

We assume that input $u_1(t)$ is available. Let the output equation for the first model be:

$$y_1(t) = g_1(x_1(t), u_1(t)) \tag{5.13}$$

For notational compactness we drop the time-variant notation from here on. The models are connected via the equation:

$$u_2(t) = y_1(t) \tag{5.14}$$

and the output equation for the second model is:

$$y_2(t) = g_2(x_2(t), u_2(t)) \tag{5.15}$$

These three equations can be computed in the presented order, which closely reflects the block diagram. The reader can verify that any other order would lead to forward references. For an off-line software implementation, we elaborate on Fig. 5.1, resulting in Fig. 5.5. As in Fig. 5.1, computation of the state prediction needs to precede the state and time updates. This represents the only constraint on the order of these equations, which can be grouped per model (as in Fig. 5.5) or per equation type.

[1] Otherwise a new single model with a state variable vector resulting from the concatenation of $x_1(t)$ and $x_2(t)$ could be derived.

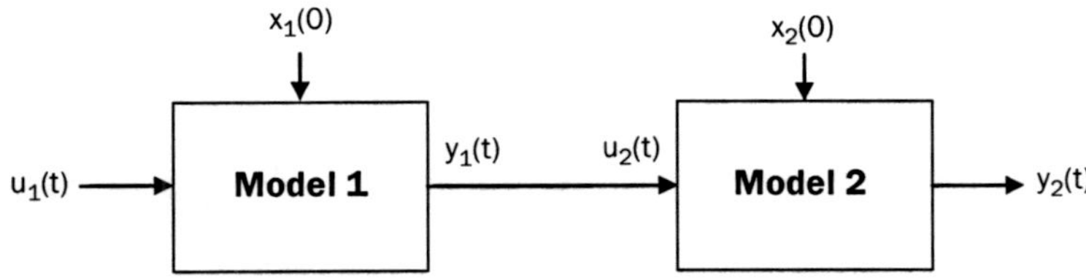

Figure 5.4 Simplified block diagram of two connected models.

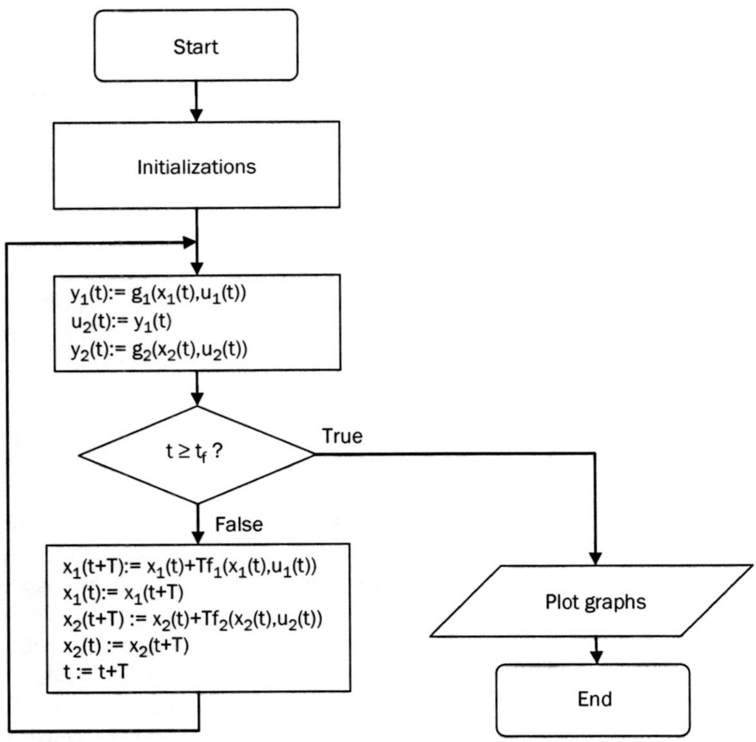

Figure 5.5 Flow diagram for off-line software implementation of in series state variable models.

This algorithm can easily be generalized to more than two models in series. Now let us assume that there is feedback from Model 2 to Model 1, maintaining the simplification concerning the absence of external input or output (Fig. 5.6).

For example, consider the cardiorespiratory system and control of breathing (Fig. 3.6). We add the connecting equation

$$u_1(t) = y_2(t) \tag{5.16}$$

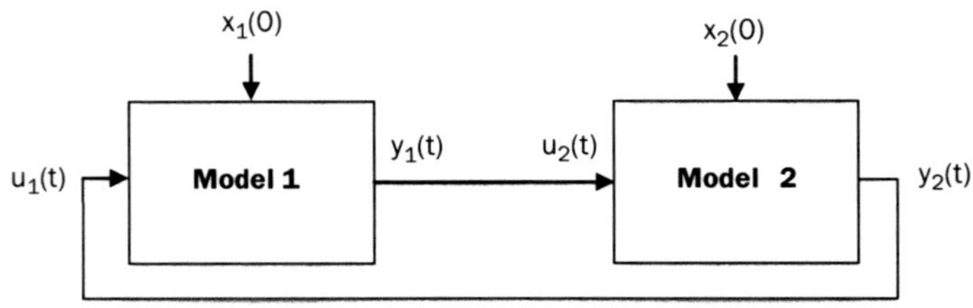

FIGURE 5.6 Simplified block diagram of two connected models with feedback.

Equations (5.13)–(5.16) cannot be computed in the presented (or any other) order, because of a circular reference. For example, substituting Eq. (5.16) in Eq. (5.13), Eq. (5.13) in Eq. (5.14), and Eq. (5.14) in Eq. (5.15), we find

$$y_2(t) = g_2(x_2(t), g_1(x_1(t), y_2(t))) \tag{5.17}$$

with a circular reference to $y_2(t)$. If such circular references occur, they are often the result of a modeling error. To avoid them in the presented example, at least one of the models should not have immediate throughput from the input $u(t)$ to the output $y(t)$, for example, Model 1:

$$y_1(t) = g_1(x_1(t)) \tag{5.18}$$

Output and connecting equations can now be computed in an order that avoids forward and circular references and that still closely reflects the block diagram (Fig. 5.7). The state predictions and state and time updates are the same as those in Fig. 5.5.

Elaborating on the feedback example, we now consider the case in which Model 2 uses an integration step size T_2, which is an integer multiple of T_1; that is, the equations of Model 2 can (and should) be computed at a lower frequency than those of Model 1. This would be the case for the chosen example: the cardiorespiratory system and control of breathing (Fig. 3.6), where the control of breathing model would only have to execute once per respiratory cycle. Note that T_2 can be variable. The algorithm needs to verify whether $t - t_0 = nT_1$ is a multiple of T_2, before computing Model 2 equations (Fig. 5.8).

Going from the high-frequency output $y_1(t)$ to the low-frequency input $u_2(t)$ involves sampling or averaging. Going from the low-frequency output $y_2(t)$ to the high-frequency input $u_1(t)$ involves interpolation, for example, zero-order hold or linear interpolation. Also note the use of T_2 in the state prediction for Model 2. In this algorithm there is no specific $t := t + T_2$ assignment and the discrete times n and m are not explicitly maintained. We further observe that

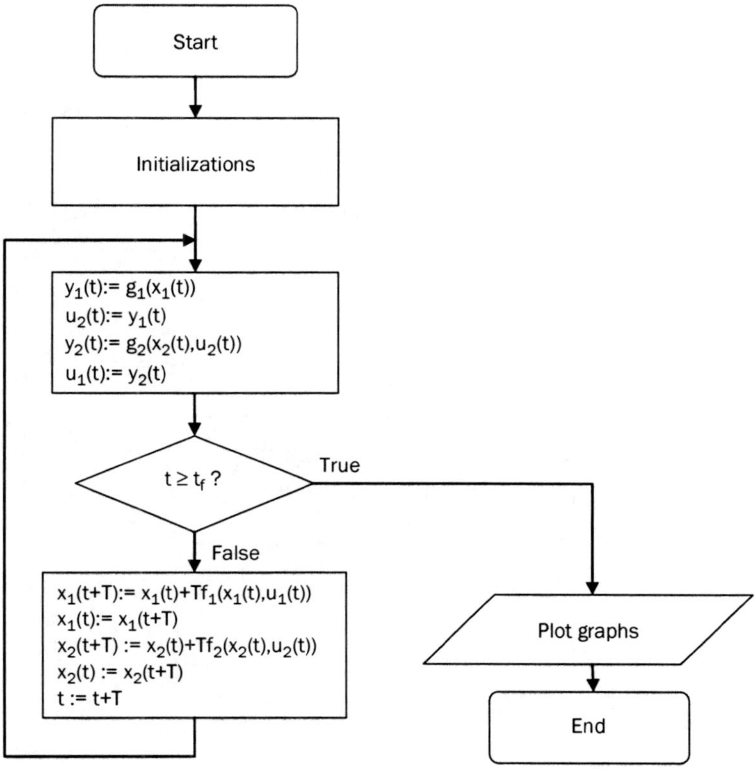

FIGURE 5.7 Flow diagram for off-line software implementation of state variable models with feedback.

the change from identical to different sampling intervals is not visible in the block diagram (Fig. 5.6) but has a considerable effect on the flow diagrams (Figs. 5.7 and 5.8). In the algorithm of Fig. 5.8, code for the two models is divided over five distinct segments, which is not conducive to modularity. This algorithm may be simplified in several situations. For example, when Model 2 is static, there is no state prediction for that model, and the second T_2 branching disappears.

We remind the reader of two simplifying assumptions applied in this section: the absence of external inputs and outputs in all three considered cases, and the absence of immediate throughput in Model 1 for the last two. Absence of immediate throughput in Model 2 leads to fairly trivial changes in the equations and algorithms. Generalization to the case including external inputs and outputs requires careful writing out of the continuous-time equations corresponding to the block diagram of Fig. 5.9.

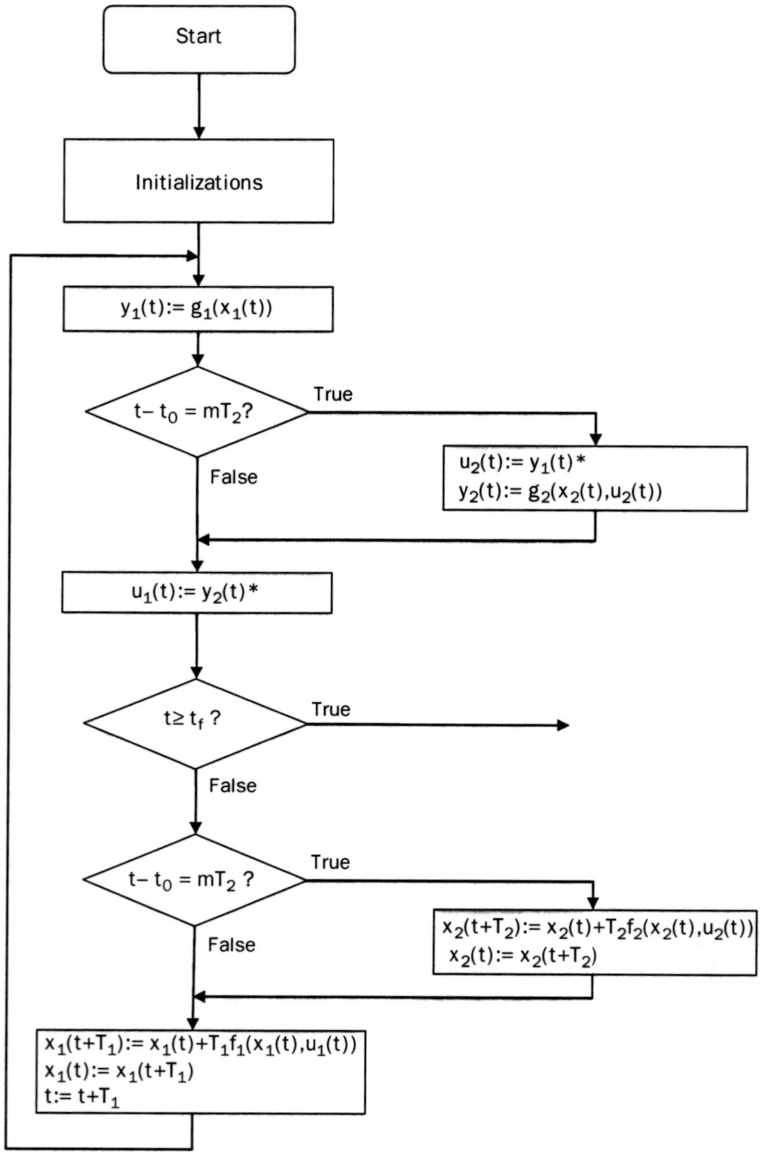

FIGURE 5.8 Flow diagram for off-line software implementation of state variable models with feedback and different integration step sizes. Plot graphs and algorithm end as in Fig. 5.1. *see main text.

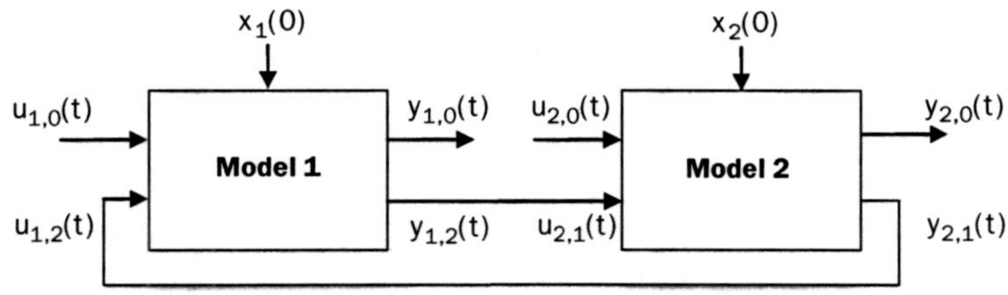

FIGURE 5.9 Block diagram of two connected models with feedback.

For example, for Model 1:

$$\dot{x}_1(t) = f_1(x_1(t), u_1(t)) \tag{5.19}$$

where:

$$u_1(t) = \begin{pmatrix} u_{1,0}(t) \\ u_{1,2}(t) \end{pmatrix} \tag{5.20}$$

and:

$$y_{1,0}(t) = g_{1,0}(x(t), u_1(t)) \tag{5.21}$$

$$y_{1,2}(t) = g_{1,2}(x(t), u_1(t)) \tag{5.22}$$

paying close attention to potential immediate throughput from the $u_{1,2}(t)$ part of $u_1(t)$ to $y_{1,2}(t)$. Then the corresponding discrete-time equations need to be added to appropriate locations in the algorithm, in terms of references and calling frequencies.

Review Problems

5.1 Select an algorithm for a code implementation for validation of the mathematical model of the uncontrolled cardiovascular subsystem, derived as part of Review Problem 4.1. Does the software have to run in real time? Make sure that the user interface allows you to reproduce experiments and present data reflecting the validation data. Implement this algorithm in a programming language of your choice. Verify the code and its operation.

5.2 Discretize, program, and simulate the state variable model of Sec. 4.2, using the algorithm of Fig. 5.1. Parameters: $R = 1\ \Omega$, $L = 1$ H, $C = 1$ F. Initial state variables: $v_c(t_0) = i_l(t_0) = 0$. Input: $e(t) = 1$ V for $0 < t < 20$ s, 0 V otherwise. Plot the output variable $v_l(t)$. How did you select the integration step size T? Experiment with parameter variations; predict the expected changes in response first.

References and Further Reading

Chapra SC, Canale RP. *Numerical Methods for Engineers*, 6th ed. New York: McGraw-Hill; 2010.

Ljung L. *System identification, theory for the user*, 2nd ed. Upper Saddle River, NJ: Prentice Hall; 1999.

CHAPTER 6

Simulation Results and Model Validation

Model verification targets several, relatively simple aspects of model quality. A model is considered conceptually valid if incorporated theories, assumptions, and data are reasonable for the intended purpose. A model is considered operationally valid for a set of experimental conditions if its accuracy is acceptable for the intended purpose. Measures of accuracy can be based on comparison of time or frequency domain features of system and model outputs. Operational validity can be based on target data obtained on human subjects or from animal experiments, or by presenting simulation results to clinical experts. The used evaluation method depends greatly on the intended purpose and available resources.

6.1 Definitions and Overall Procedure

In this chapter we consider the quality of a model used in simulation. *Model* is used here in a generic sense, including conceptual and mathematical models, code implementation, and simulation results. We distinguish between to *verify*—"Conclusively demonstrate by presentation of facts or by sound reasoning or argument"—and to *validate*—"Corroborate or support on a sound basis or authority" (Webster's third new international dictionary, 2002). *Model verification* targets several, relatively simple aspects of model quality, such as inclusion of all required input and output variables, correct derivation of model equations based on a conceptual model and physical laws, and correct implementation of model equations in code. Verification leads to unambiguous, factual results, often in the form of a yes or no answer. *Model validation* targets other, more complex aspects of model quality. For example, are the simplifying assumptions reflected in the structure of a conceptual model justified in view of the intended purpose? Validation often refers to more or less subjective observations by experts, and is by nature a gradual notion; we may speak of degrees of validity. For model validation, we further distinguish the quality of what was put into a model and the quality of its behavior:

- A model is considered *conceptually valid* if incorporated theories, assumptions, and data are reasonable for the intended purpose.
- A model is considered *operationally valid* for a set of experimental conditions if its accuracy is acceptable for the intended purpose.

A central role in all considerations on verification and validation is played by the "intended purpose" as reflected in the model requirements. In these definitions, the subjective aspect of validity becomes apparent in the use of the words "considered," "reasonable," and "acceptable" (by whom, to whom, according to which criteria?). Both definitions refer to a single model; multiple coupled models should, as much as possible, be validated in isolation first. Conceptual validity mainly concerns the conceptual and mathematical models, whereas operational validity refers to simulation results and target data. Therefore, the basis for conceptual validity is laid during the modeling process, while operational validity is established at the end of the modeling process, with a completed model, before it is used in practical simulations. Conceptual validity is often established by model developers, whereas operational validity is increased when carried out (or at least verified) by independent evaluators or model users. Input from physiologists and clinicians may be of use in establishing both conceptual and operational validity, but a natural focus would be that of physiologists on conceptual matters and of clinicians on operational matters. Model verification and conceptual validity involve meeting requirements, as presented in the previous chapters. The focus of this chapter is on simulation results and operational validity. In subsequent sections we elaborate on "accuracy," which refers to a quantitative or qualitative comparison between target data and simulation results, and on the "set of experimental conditions," which points to a necessarily limited, range of model validity.

Documentation is central to model verification and validation. If models are evaluated by professionals who are not part of the development team, and possibly do not have a modeling background, then a summary of model attributes and a critical assessment of those with respect to the model requirements, as formulated in Chap. 2, greatly contributes to model verification and validation. Table 6.1 lists a number of questions that can be addressed in such a summary. Some questions on operational validation will be further clarified in subsequent sections. This list refers to a newly developed model. If a previously developed model is reused, for example, for a different patient population, several verification and validation questions may

Verification	
Were all required input and output variables included?	
Do they have appropriate units, ranges, resolution, and bandwidth?	
Are the required anatomical structures and physiological processes included in the conceptual model?	
Was correct derivation and implementation of model equations verified?	
Was adequacy of integration method and step size verified? This may involve a range of parameter values.	
Do simulation results match known analytical solutions or steady-state results?	
Does the model code run fast enough for the envisioned application?	
Was the model adequately verified?	☐
Conceptual validation	
What simplifying assumptions were made concerning structure, and what was their justification?	
What simplifying assumptions were made concerning function, and what was their justification?	
Does the level of representation detail match the objectives?	
What underlying laws and principles were used, and was their use appropriate?	
What assumptions were made in deriving the mathematical model, and what was their justification?	
What assumptions were made for parameter estimation, and what was their justification?	
What data were used in parameter estimation, and were the data appropriate in view of the objectives?	
Are the resulting parameter values physiologically plausible?	
Were physiologists knowledgeable in the domain of the model consulted? What was their opinion?	
Is the model considered conceptually valid?	☐
Operational validation	
What are the patients, pathologies, conditions, incidents, and interventions the model needs to be validated for?	
Are target data on human subjects available for these situations?	
If not: are target data on animal subjects available for these situations, and can the data be scaled?	
Were clinical experts who have been exposed to these situations consulted? What was their opinion?	

Table 6.1 Questions Concerning Physiologic Model Verification and Validation

Do experimental conditions of the model reflect those of the target data?	
What measure of accuracy was used?	
Are the results considered acceptable in view of the project objectives?	
Is the model considered operationally valid for the listed situations?	
Was model behavior explored beyond the available target data? Was it considered plausible?	☐

Table 6.1 Questions Concerning Physiologic Model Verification and Validation (*continued*)

have already been answered. In general, the list should be amended based on specific project needs.

This rather extensive list shows that validity has a cost. Depending on the phase of model development or use, increasing levels of validity may be required, and certain elements of the list may be revisited, or the range of validity may be expanded, for example, via model modifications enhancing conceptual validity, or inclusion of additional experimental conditions broadening the range of operational validity. A cost–benefit analysis could consider development or evaluation costs and the value of a sought-after increment in validity.

A physiologic model is rarely designed and evaluated in isolation; its requirements are embedded in more general project goals, and evaluation of conceptual and operational validity refers back to these requirements. Additional quality questions arise around the use of a model in a simulator. For example, modularity and interaction with other functional units, such as other models or simulator hardware and user interfaces, become important. As argued in Chap. 2, the quality of such interfaces will influence the overall impact of a model-driven simulator. In Chap. 12 we will explore this broader perspective further.

With time and use of a simulator, model applications may evolve beyond the originally intended purpose, and model and simulator validation in new experimental conditions may be necessary and can lead to modifications. A dramatic illustration of this observation is the use of a model-driven simulator, designed and developed between 1986 and 1996 for training of anesthesia residents, for totally unanticipated post 9/11 disaster medicine drills. For additional information on verification and validation of simulation models we point to Sargent (2007).

6.2 Quantitative and Qualitative Methods for Establishing Accuracy

The principle of establishing operational validity is visualized in Fig. 6.1.

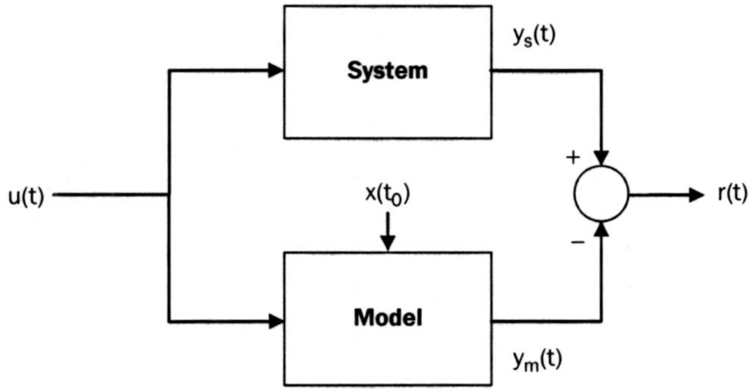

Figure 6.1 Principle of establishing operational validity.

The system and its model are assumed to be in identical initial conditions. Then the same input is applied to both, the model output $y_m(t)$ is subtracted from the system output $y_s(t)$, and the residual $r(t)$ is evaluated.[1] The target data $[u(t),y_s(t)]$ should not have been used previously for model parameter estimation. The analysis of the residual can be quantitative of qualitative. The residual can be processed to lead to a measure of accuracy A, for example, for a continuous-time system with a single output

$$A_1 = \frac{1}{T}\int_0^T \left|y_s(t) - y_m(t)\right| dt \tag{6.1}$$

$$A_2 = \frac{1}{T}\int_0^T \left(y_s(t) - y_m(t)\right)^2 dt \tag{6.2}$$

$$A_3 = \frac{1}{T}\int_0^T \frac{\left|y_s(t) - y_m(t)\right|}{\left|y_s(t)\right|} dt \tag{6.3}$$

representing, respectively, the integral of absolute residuals, the integral of squared residuals, and the integral of normalized absolute residuals. T represents here the total record length of the evaluated data. A_3 applies only if $y_s(t)$ does not have zero values. These measures can easily be adapted to multivariable and discrete-time systems. Other measures of accuracy can be based on comparison of overall time or frequency domain features of system and model outputs,

[1] We reserve the term "error" for use in feedback control. Note that "residual" is used here in a deterministic sense.

such as autocorrelation and frequency response. Specific features of the output signals, for example, amplitude, baseline, time of peak effect, and bandwidth, can also be used in accuracy measures. Simulated waveforms are more easily evaluated visually, by domain experts. Such evaluations are qualitative rather than quantitative, although semi-quantitative resemblance scales can be established, with or without explicit comparison to real system outputs. For examples of the latter, see Bastos (2011). Target data can also be obtained from a more complete, previously validated model. The used evaluation method depends greatly on the intended purpose and on available resources.

6.3 Range of Validity, Target Data, and Experimental Conditions

As mentioned in Sec. 6.1, the range of operational validity of a model is determined by the set of experimental conditions for which it is validated. These experimental conditions should match the population, pathologies, conditions, incidents, and interventions part of the requirements (Tables 2.1 and 6.1). Often, our purpose is to reflect a plausible patient, within normal interpatient variability (also see Table 2.3). Some of these target data will reflect steady states, whereas others will involve dynamic experiments. A fundamental paradox of modeling and simulation, and not only in biomedical engineering, is that we often want to explore unknown and potentially extreme, rare, and dangerous system behavior. Target data for those situations, obtained in carefully controlled conditions, are not always available. Another paradox is that we often use a model to represent dynamic systems with one or more continuous and possibly interacting independent variables, Sec. 2.1. The number of possible combinations of input variables and responses is very large and not all combinations can be expected to be validated.

If our model is for human subjects, and if no target data are available for that population, then data from animal experiments are sometimes used. There are ethical concerns with carrying out animal experiments, and even with using data from such experiments. Moreover, validation is complicated by the fact that model inputs and outputs have to be scaled between species, or two sets of model parameters are required: one for the modeled human, and one for the species of the target data. Model validity is not established as clearly as when using human target data. A third option for model validation is to present simulation results, combined with carefully formulated questions, to clinical experts (also see the previous section). They may not have had much exposure to rare, extreme situations, and do not always agree. We frequently use combined strategies for model validation (see, for example, Sá Couto, 2002; Bastos, 2011). As briefly mentioned in the previous section, data used for validation need to be independent from those used in parameter estimation, otherwise we could simply be verifying that these data were correctly incorporated into the model. We can distinguish an identification set and a validation set.

The experimental conditions of the simulation model need to be matched with those of the system yielding the target data. For example, an initial state involving hypovolemia may influence the response of a patient to mechanical ventilation. Model behavior can be explored beyond the available target data, and may yield useful validity information. One example is a sensitivity analysis, where one input or one parameter is varied at a time, and the effect on output variables is observed. A sensitivity analysis may provide information about linearity, assist in parameter estimation, and provide general indicators on model behavior. We will explore this method further in Chap. 11. It is also useful to evaluate the model response in extreme conditions, even if no target data are available. This helps in assuring that the model correctly implements boundary conditions, for example, blood volume should not become negative on blood loss.

Review Problems

6.1 Compare the simulation results obtained with the code developed as part of Review Problem 5.1 to validation data. Which comparison method(s) did you use? What are the ranges of the input variables for which you have reliable validation data? Comment on conceptual and operational validity of the developed model. Does the model need to be improved? How could that be accomplished? What are the steps required to complete the model and the code, and to transform it into a practical simulator for use in medical education?

6.2 Can a model be conceptually valid, but not operationally valid, and vice versa? What would cause such discrepancies?

References and Further Reading

Bastos ML. Mathematical models for educational simulation of uterine contractions during labor [thesis]. Portugal: University of Porto; 2011.

Sá Couto PM, van Meurs WL, Bernardes JF, Marques de Sá JP, Goodwin JA. Mathematical model for educational simulation of the oxygen delivery to the fetus. *Control Eng Practice.* 2002; 10:59–66.

Sargent RG. Verification and validation of simulation models, in Henderson SG, et al. (eds): Proceedings of the 2007 Winter Simulation Conference, pp 124–137.

Zijlmans M, Sá-Couto CD, van Meurs WL, Goodwin JA, Andriessen P. Corrected and improved model for educational simulation of neonatal cardiovascular pathophysiology [technical report]. *Simul Healthc.* Spring 2009; 4(1):49–53.

PART II

Applications

CHAPTER 7

A Model of the Cardiorespiratory System

In this chapter we present a framework for modeling projects in cardiorespiratory physiology. Qualitative requirements for our model of the whole cardiorespiratory system include simulation of a number of clinically monitored physiologic variables, and simulation of appropriate reactions to the administration of fluid and to re-breathing of carbon dioxide. We derive an overall conceptual model in the form of a block diagram of the subsystems: hemodynamics, lung mechanics, uptake and distribution, control of circulation, and control of breathing.

In this part of the book we apply the theory introduced in Part I. The goals of Part II are to further illustrate and expand presented concepts and procedures and to provide the reader with a consistent library of conceptual and mathematical models of the normal cardiorespiratory system. As mentioned in the foreword, they represent a simplified version of the cardiorespiratory models used in the Human Patient Simulator (HPS®) and derived products (van Meurs, 1997, www.meti.com). These models will be of intermediate complexity, so that they can be either simplified or expanded for other applications. We follow the top-down, goal-oriented approach to modeling presented in Part I. In the subsequent sections, we formulate qualitative requirements for a model of the whole cardiorespiratory system and derive an overall conceptual model in the form of a block diagram of subsystems, following the procedures outlined in Secs. 2.1 and 3.1, respectively. This approach highlights the interconnectedness of the cardiorespiratory system and will give the reader a framework for future modeling projects in this area. In the subsequent chapters, the conceptual model of the cardiorespiratory system is deepened by modeling the circulation via fluid circuit elements representing the heart and vessels, followed by a model for fluid mechanics of lungs and airways. Addition of compartment models of respiratory gas transport and transfer functions representing control of circulation and spontaneous breathing completes this process. Mathematical models corresponding to these conceptual models are derived. These models are all in state variable notation, so software implementation can be based on the algorithms presented in Chap. 5. Most of the presented models are referenced to the scientific literature; we point to this for validation of individual models. Full integration of all seven presented mathematical models in software and validation of integrated operation are beyond the scope of this introductory text.

7.1 Model Requirements

For information on basic cardiorespiratory physiology we point to the references in Chap. 2. Our main focus will be on normal physiology, and we expect our model to react appropriately to administration of fluid, for example, to evaluate the Frank-Starling curve of the heart, and to re-breathing of carbon dioxide to test the ventilatory response. We would like for our model to simulate a number of key physiological variables that are also monitored clinically, such as heart rate, pulsatile systemic and pulmonary blood pressure waveforms, and cardiac output. On the respiratory side, it should simulate tidal volume, respiratory rate, oxygen saturation in systemic arterial blood, and the partial pressures of oxygen and carbon dioxide in systemic arterial and venous blood. Table 7.1 and Fig. 7.1 list and complete these requirements.

These requirements contain several simplifying assumptions. For example, the electrocardiogram (ECG), the plethysmogram,[1] airway pressures, blood pH, and hematocrit[2] are not considered output variables and do not have to be modeled as such. In some cases this means that the underlying structures and functions do not have to be represented in the model, for example, electrophysiology underlying the ECG. In other cases, it only means that variables are assumed constant or that their evolution will not be displayed as an output variable.

To be able to simulate the interaction between these variables, internal structures and functions that should be incorporated include the basic fluid dynamics of the heart and systemic and pulmonary vessels; basic airway, lung, and chest wall mechanics; pulmonary uptake and distribution of the respiratory gases oxygen and carbon dioxide; and basic representations of the physiologic control mechanisms of circulation and spontaneous breathing. We will focus on short-term phenomena, roughly on a time scale from 2 to 300s, ruling out, on the one hand, fast phenomena like beat-to-beat heart

[1] The plethysmogram is the waveform of the pulse oximeter.

[2] Hematocrit is the fraction of whole blood that is occupied by red blood cells.

1. Organ or physiological system	Cardiorespiratory system
2. Population	Healthy adult male
3. Physiological state of the simulate subject	Normal baseline
4. Pathologies	None
5. Monitored variables and clinical signs	See main text
6. Critical incidents	None
7. Interventions	Fluid administration, re-breathing of CO_2
8. Overall block diagram	Fig. 7.1
9. Internal structures and functions	See main text

Table 7.1 Qualitative Model Requirements

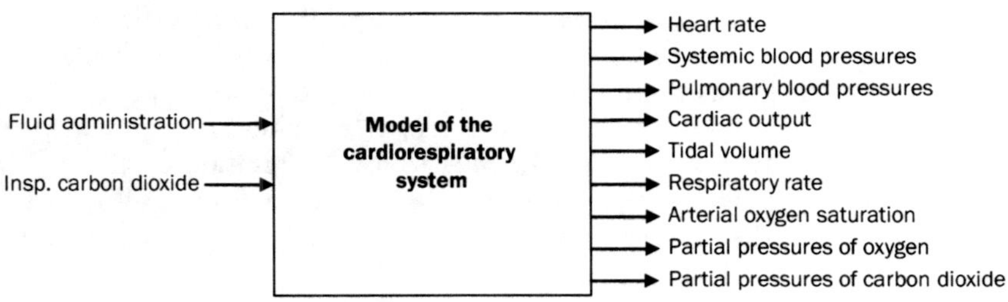

Figure 7.1 Overall block diagram of the cardiorespiratory model.

rate variability, and on the other, slower dynamics of fluid balance and endocrine aspects of physiologic control mechanisms.

7.2 Conceptual Model

To come to a conceptual model of this rather complex system, we take as our starting point the graphical representation of the cardiorespiratory system (Fig. 3.2), to which we add the control of circulation.[3] We observe that the system involves volume, concentration, and action-potential-type variables. Respiratory gases are transported by both blood and gas. Control systems affect the

[3] At this stage of modeling, a search for published models would be warranted, and lead to many results, but for didactic reasons we will build the conceptual model from the ground up.

circulatory and respiratory systems. Using these criteria, the identified internal structures and functions can be divided in causally connected subsystems as in the block diagram of Fig. 7.2. The circulation of blood is influenced by ventilation via the intrapleural or intrathoracic pressure. Control of breathing mainly depends on blood gases. Control of circulation depends on those, as well as on blood pressures. However, the influences of blood gases on the circulation only play a minor role in the interventions that we would like to simulate, so they will not be considered further. For this system, and at this level of conceptual modeling, separation along anatomical lines, also indicated in the block diagram, plays a more limited role. To keep oversight, only connecting variables are indicated in the diagram, and are given generic names. Table 7.2 associates the input and output variables of Fig. 7.1 with the subsystems of Fig. 7.2.

By considering fluid administration as an input to the hemodynamic model, we make the simplifying assumption that the fluid does not affect the oxygen carrying capacity of the blood, part of the uptake and distribution model. The input and output variables of the overall block diagram are measurable by design, per the requirement on clinically monitored variables. Specific instantiations of the generic variables included in the block diagram may not always be measureable. For example, depending on the choice of a specific structure for the uptake and distribution model, the blood flow rates and volumes may not be directly measurable. Activities of the sympathetic and parasympathetic nervous systems are useful concepts when discussing cardiorespiratory control, and their

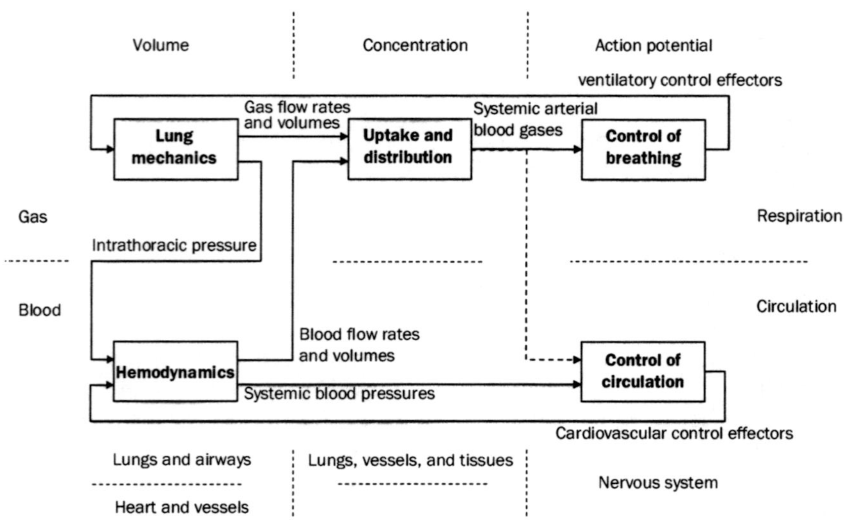

FIGURE 7.2 Block diagram of the cardiorespiratory system and subsystem separation criteria (dashed lines).

Overall inputs	To
Fluid administration	HD
Inspired carbon dioxide	UD
Overall outputs	**From**
Heart rate	CC, together with other control effectors
Systemic blood pressures	HD
Pulmonary blood pressures	HD
Cardiac output	HD, together with other blood flow rates
Tidal volume	LM, implicit in gas flow rates and volumes
Respiratory rate	CB, together with other control effector
Arterial oxygen saturation	UD
Partial pressures of oxygen	UD
Partial pressures of carbon dioxide	UD

HD, hemodynamics; UD, uptake and distribution; CC, control of circulation; LM, lung mechanics; CB, control of breathing.

TABLE 7.2 Overall Input-Output Variables and Associated Subsystems

representation is essential when modeling relatively fast phenomena like beat-to-beat heart rate variability, but this is not a requirement for the present model. The block diagram clearly divides the circulatory and respiratory loops, and shows the main connection between them, namely, hemodynamics driving pulmonary uptake of respiratory gases and their distribution via the circulation. It also shows secondary connections, with intrathoracic pressure influencing the circulation and arterial blood gases influencing control of circulation. Structures and functions that are not included in the block diagram, besides the already mentioned electrophysiology of the heart, are, for example, a fluid balance, the effect of lung stretch receptors on control of breathing, a myocardial oxygen balance, and local control of vascular beds. If, however, inclusion of such systems were part of the requirements, the block diagram could easily be expanded to accommodate that. A quite similar conceptual model (van Meurs, 1997) needed only minor adaptations to also model drug influences on the cardiorespiratory system. In the subsequent chapters we will present conceptual and mathematical models for all five blocks of Fig. 7.2.

Reference and Further Reading

Van Meurs WL, Good ML, Lampotang S. Functional anatomy of full-scale patient simulators. *J Clin Monit* Sep1997; 13(5):317-24.

CHAPTER 8
Circulation

In this chapter we further specify the requirements of the pulsatile hemodynamic model. We introduce a conceptual model consisting of an elastance generator and a model for basic hemodynamics. The latter is represented by a hydraulic component diagram. Complete mathematical models for both subsystems are presented. The model for the basic hemodynamics is shown to have a state variable form.

8.1 Model Requirements

From Fig. 7.2 and Table 7.2 we obtain the overall block diagram of the hemodynamic model shown in Fig. 8.1.

Table 8.1 further specifies the input and output variables. In this part of the book, we focus on conceptual and mathematical models. Therefore, indications of range, resolution, and bandwidth, which play an important role in code implementation and interfacing, are omitted.

Fluid administration and intrathoracic pressure are given in typical units used in clinical practice and modeling of lung mechanics, respectively. We model control of circulation via four effector variables: heart rate, myocardial contractility, systemic arteriolar resistance, and venous unstressed volume. This choice includes the assumption that, for the intended purpose, control of circulation can be represented via relatively easily observable physiologic variables, without a more detailed representation of the autonomic nervous system. Myocardial contractility is given in (maximum) elastance units (also see Sec. 4.4). Heart rate and myocardial contractility, as defined above, only have a value once per cardiac cycle. This is not the case for the two additional control effectors, which are

FIGURE 8.1 Overall block diagram of the hemodynamic model.

Input variable	Symbol	Unit	Origin
Fluid administration	$f_{fa}(t)$	ml/min	External input
Intrathoracic pressure	$p_{it}(t)$	cm H_2O	Lung mechanics
Heart rate	$hr(n)$	min^{-1}	Control of circulation
Myocardial contractility	$mc(n)$	mm Hg/ml	Control of circulation
Systemic arteriolar resistance	$r_{sart}(n)$	mm Hg ml^{-1} min	Control of circulation
Venous unstressed volume	$v_{vu}(n)$	ml	Control of circulation
Output variable	**Symbol**	**Unit**	**Destination**
Mean arterial blood pressure	$map(n)$	mm Hg	Control of circulation
Systemic arterial blood pressure	$p_s(t)$	mm Hg	External
Central venous blood pressure	$p_{cv}(t)$	mm Hg	External
Pulmonary arterial blood pressure	$p_p(t)$	mm Hg	External
Pulmonary capillary wedge pressure	$p_{pcw}(t)$	mm Hg	External
Cardiac output	$co(k)$	L/min	External
Pulmonary tissues blood flow rate	$f_{pt}(t)$	ml/s	Uptake and distribution
Arterial pool blood flow rate	$f_{ap}(t)$	ml/s	Uptake and distribution
Systemic tissues blood flow rate	$f_{st}(t)$	ml/s	Uptake and distribution
Venous pool blood flow rate	$f_{vp}(t)$	ml/s	Uptake and distribution
Arterial pool blood volume	$v_{ap}(t)$	ml	Uptake and distribution
Systemic tissues blood volume	$v_{st}(t)$	ml	Uptake and distribution
Venous pool blood volume	$v_{vp}(t)$	ml	Uptake and distribution

Here *n* represents the cardiac cycle, and *k* intermittent sampling to obtain cardiac output.

TABLE 8.1 Input and Output Variables of the Hemodynamic Model

continuous-time variables, but defining them as discrete-time variables avoids a hybrid formalism of the control of circulation model, also facilitating software implementation. We further assume that the only hemodynamic variable that affects control of circulation is mean arterial blood pressure, which is also available once per cardiac cycle. In Chap. 10 we will come back to these assumptions, and provide references to the control of circulation model. The output pressures correspond to clinically monitored waveforms. Central venous pressure can be measured in various locations, but is defined as right atrial pressure. Pulmonary capillary wedge pressure is measured via inflation of a small balloon at the tip of a catheter blocking one of the pulmonary arteries. Via a static fluid column, the pressure in the pulmonary veins is measured. Cardiac output is sampled intermittently, for example, via measuring blood temperature in the pulmonary arteries, after upstream injection of a bolus of cold fluid. The thermodilution gives an indication of flow rate. It is given in clinical units. Anticipating the design of the uptake and distribution model (Chap. 9), we consider that, for the purpose of our model, respiratory gas exchange can be adequately reflected via four perfused homogeneous compartments: pulmonary tissues and alveolar space, arterial pool, systemic tissues, and venous pool. Gas content in pulmonary blood will be ignored. These considerations result in the listed output variables. These flow rates and volumes are rarely monitored as such, so we assign units that are convenient for modeling. In Chap. 9 we will come back to the use of pulsatile versus average flow rates and volumes.

Only compound pulmonary and systemic flow rates are required; these blood flows are not divided over separate left and right lungs, or over different organs or tissue groups. This requirement makes it likely that the model will not be designed to include such flow rates either. If flow rates and blood volumes in specific organs or tissue groups can be considered a fixed fraction of total flow rate and volume, then they can be computed as such. If not, for example, when modeling local control of coronary or brain circulation, the requirements need to be expanded, and additional organs or tissue groups need to be added.

8.2 Conceptual Models

A relatively simple pulsatile hemodynamic model that meets most of the input–output requirements was among several conceptual and mathematical models presented by Beneken (1965). More recently it was described and used in a new application by Goodwin et al. (2004). This model separates the electromechanical processes, which are responsible for the generation of the time-varying elastances of the heart chambers, from the basic hemodynamics (Fig. 8.2).

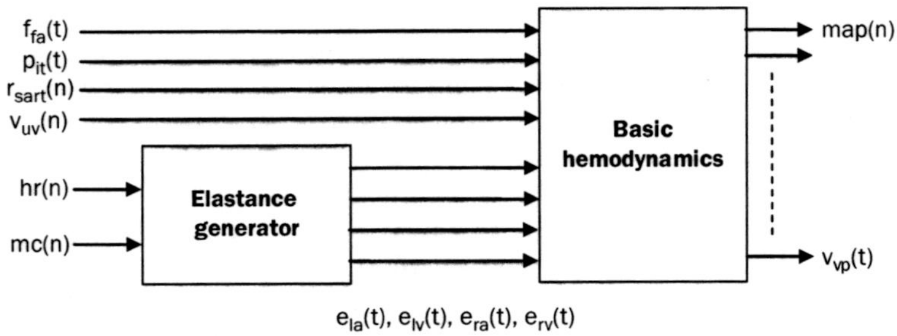

FIGURE 8.2 Block diagram of the hemodynamic model. See Table 8.1 for output variables.

An empirical elastance generator uses the input variables heart rate and myocardial contractility, which are available once per cardiac cycle, to parameterize the continuous-time elastances for all four heart chambers (also see Sec. 4.4). Figure 8.3 shows a component diagram for the basic hemodynamics, with some minor adaptations with respect to the Beneken model, using the circuit elements of Table 4.6.

The adaptations concern the separation of the systemic tissue resistance in a variable arteriolar and a fixed venular part, and the addition of a compliance containing blood volume in the systemic tissues. The first adaptation provides an explicit element for the input variable: systemic arteriolar resistance. The second adaptation facilitates coupling the hemodynamic model with the uptake and distribution model described in the next chapter. Other input variables affect this model as follows: Fluid can be administered intravenously via a catheter in the extrathoracic systemic veins. The intrathoracic pressure is considered an external pressure for all compliances, with the exception of the systemic arteries, systemic tissues, and extrathoracic systemic veins. Venous unstressed volume is distributed over the extrathoracic and intrathoracic veins. Output variables include the mean arterial blood pressure, which is the average of the pressure in the aorta.[1] The other blood pressure waveforms are also associated with specific compliances in the conceptual model. Cardiac output is the average of the flow rate in the pulmonary arteries, or can be computed as the product of heart rate and stroke volume. Table 8.2 lists the homogeneous compartments of the uptake and distribution model and the corresponding compliances of the hemodynamic model.

A compartment volume output of the hemodynamic model is simply the sum of compliance volumes corresponding to each

[1] We have argued in Goodwin (2004) and Zijlmans (2009) that it is more correct to consider the elements named aorta and systemic arteries in Beneken(1965) a low-order empirical approximation of the whole systemic arterial bed, with a single arterial pressure waveform output $p_{ao}(t)$.

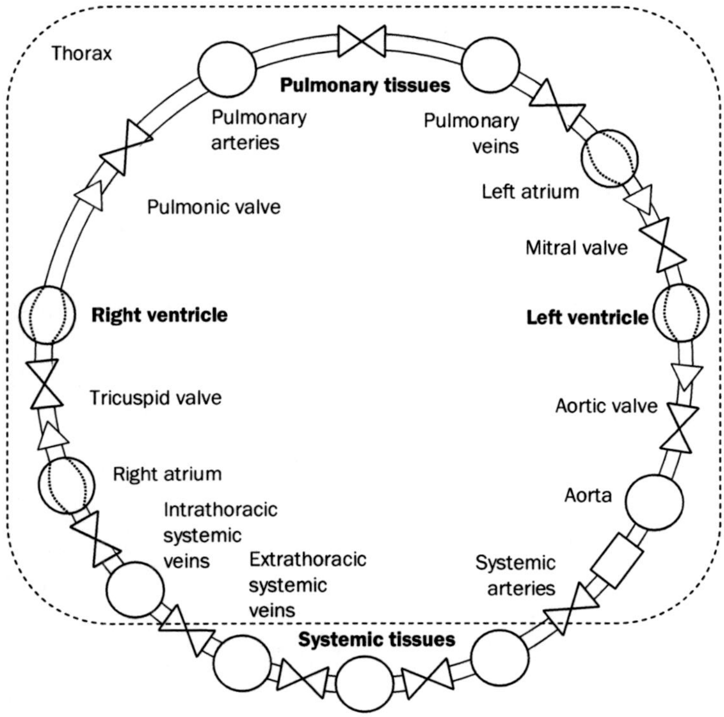

FIGURE 8.3 Conceptual model of basic hemodynamics.

Compartment	Compliances
Pulmonary tissues	None
Arterial pool	Pulmonary veins, left atrium, left ventricle, aorta, systemic arteries
Systemic tissues	Systemic tissues
Venous pool	Extrathoracic veins, intrathoracic veins, right atrium, right ventricle, pulmonary arteries

TABLE 8.2 Homogenous Compartments of the Uptake and Distribution Model and Compliances in the Hemodynamic Model

compartment. The compartment flow rate outputs of the hemodynamic model correspond to the inflow rates of each compartment. For example, the arterial pool blood flow rate is the flow rate into the pulmonary veins.

Simplifying assumptions of this conceptual model are that inertial phenomena are only represented in the aorta (where the acceleration of blood is the highest). Simplifying assumptions made at an earlier stage, for example, concerning the undivided systemic and pulmonary blood flow rates, become more visible in the component diagram.

8.3 Mathematical Models

Elaborating on the parameterized elastance curve presented in Eq. (4.32), the time-varying elastances of the four heart chambers can be represented as follows:

$$a_a(t) = \begin{cases} \sin(\pi t / T_{as}) & for \quad 0 < t < T_{as} \\ 0 & (otherwise) \end{cases} \tag{8.1}$$

$$e_{la}(t) = E_{lamin} + (E_{lamax} - E_{lamin})a_a(t) \tag{8.2}$$

$$e_{ra}(t) = E_{ramin} + (E_{ramax} - E_{lamin})a_a(t) \tag{8.3}$$

$$a_v(t) = \begin{Bmatrix} \sin[\pi(t - T_{as} - T_{av})/T_{vs}] & for \quad T_{as} + T_{av} < t < T_{as} + T_{av} + T_{vs} \\ 0 & (otherwise) \end{Bmatrix} \tag{8.4}$$

$$e_{lv}(t) = E_{lvmin} + (E_{lvmax} - E_{lvmin})a_v(t) \tag{8.5}$$

$$e_{rv}(t) = E_{rvmin} + (E_{lvmax} - E_{rvmin})a_v(t) \tag{8.6}$$

where $a_a(t)$ and $a_v(t)$ are the atrial and ventricular activation functions, respectively.[2] The timing parameters T_{as}, T_{av}, and T_{vs} represent the durations of atrial systole, atrioventricular (AV) node conduction, and ventricular systole, respectively. The ventricular activation function is delayed by a time T_{as} + T_{av} with respect to the atrial activation function.[3] Each heart chamber has its own elastance function, with distinct minimum and maximum elastances. The elastance functions are repeated with heart period HP. The time in the cardiac cycle needs to be maintained in a state variable. This can take place inside the elastance generator (Fig. 8.2) or at a higher level (Fig. 7.2). In the latter case, the time in the cardiac cycle becomes an input variable to the generator. The generator has no other state variables.

[2] Goodwin et al. (2004) present a slightly skewed ventricular activation function, which somewhat improves the simulated systemic arterial blood pressure waveform.

[3] Strictly speaking, conduction in the AV node starts before the end of atrium systole. This can be compensated by a smaller value of T_{av}.

Using the mathematical descriptions of the components listed in Table 4.6, and following the same procedure as for the electrical analog in Sec. 4.7, we derive a mathematical model corresponding to the conceptual model of Fig. 8.3. Considering that all volumes in the compliances are above the unstressed volumes, the transmural pressures across the compliances are:

$$p_{lv}(t) = e_{lv}(t)\left[v_{lv}(t) - UV_{lv}\right] \tag{8.7}$$

$$p_{ao}(t) = E_{ao}\left[v_{ao}(t) - UV_{ao}\right] \tag{8.8}$$

$$p_{sa}(t) = E_{sa}\left[v_{sa}(t) - UV_{sa}\right] \tag{8.9}$$

$$p_{st}(t) = E_{st}\left[v_{st}(t) - UV_{st}\right] \tag{8.10}$$

$$p_{esv}(t) = E_{esv}\left[v_{esv}(t) - UV_{esv}\right] \tag{8.11}$$

$$p_{isv}(t) = E_{isv}\left[v_{isv}(t) - UV_{isv}\right] \tag{8.12}$$

$$p_{ra}(t) = e_{ra}(t)\left[v_{ra}(t) - UV_{ra}\right] \tag{8.13}$$

$$p_{rv}(t) = e_{rv}(t)\left[v_{rv}(t) - UV_{rv}\right] \tag{8.14}$$

$$p_{pa}(t) = E_{pa}\left[v_{pa}(t) - UV_{pa}\right] \tag{8.15}$$

$$p_{pv}(t) = E_{pv}\left[v_{pv}(t) - UV_{pv}\right] \tag{8.16}$$

$$p_{la}(t) = e_{la}(t)\left[v_{la}(t) - UV_{la}\right] \tag{8.17}$$

Using the unit step function $u(x) = 0$, for $x < 0$, $u(x) = 1$ for $x \geq 0$, to describe the valves[4] and the differences in resistance to forward and reverse flow at the atrial inlets, we find the flow rates:

$$f_{lv}(t) = u\left[p_{la}(t) - p_{lv}(t)\right]\left[p_{la}(t) - p_{lv}(t)\right] / R_{mv} \tag{8.18}$$

[4] The original Beneken model allows for backflow across the aortic valve, which only closes when the pressure difference is equal to –1 mm Hg, resulting in a small dicrotic notch in the simulated systemic arterial pressure signal.

$$f_{ao}(t) = u[p_{lv}(t) - p_{ao}(t)][p_{lv}(t) - p_{ao}(t)] / R_{av} \tag{8.19}$$

$$\frac{df_{sa}(t)}{dt} = [p_{ao}(t) + P_{it} - R_{sa} f_{sa}(t) - p_{sa}(t)] / L_{sa} \tag{8.20}$$

$$f_{st}(t) = [p_{sa}(t) - p_{st}(t)] / R_{sart} \tag{8.21}$$

$$f_{esv}(t) = [p_{st}(t) - p_{esv}(t)] / R_{sven} \tag{8.22}$$

$$f_{isv}(t) = [p_{esv}(t) - p_{isv}(t) - P_{it}] / R_{sv} \tag{8.23}$$

$$\begin{aligned} f_{ra}(t) = {} & u[p_{isv}(t) - p_{ra}(t)][p_{isv}(t) - p_{ra}(t)] / R_{ra} \\ & - u[p_{ra}(t) - p_{isv}(t)][p_{ra}(t) - p_{isv}(t)] / (10 \times R_{ra}) \end{aligned} \tag{8.24}$$

$$f_{rv}(t) = u[p_{la}(t) - p_{rv}(t)][p_{ra}(t) - p_{rv}(t)] / R_{tv} \tag{8.25}$$

$$f_{pa}(t) = u[p_{rv}(t) - p_{pa}(t)][p_{rv}(t) - p_{pa}(t)] / R_{pvpa} \tag{8.26}$$

$$f_{pv}(t) = [p_{pa}(t) - p_{pv}(t)] / R_{pt} \tag{8.27}$$

$$\begin{aligned} f_{la}(t) = {} & u[p_{pv}(t) - p_{la}(t)][p_{pv}(t) - p_{la}(t)] / R_{la} \\ & - u[p_{la}(t) - p_{pv}(t)][p_{la}(t) - p_{pv}(t)] / (10 \times R_{la}) \end{aligned} \tag{8.28}$$

Eq. (8.20) results from application of the fluid mechanics equivalent of Kirchhoff's voltage law to the loop: reference pressure, intrathoracic pressure, transmural pressure aortic compliance, systemic arterial fluid inductor, systemic arterial resistance, transmural pressure systemic arteries compliance, and back to reference pressure:

$$-P_{it} - p_{ao}(t) + L_{sa} \frac{df_{sa}(t)}{dt} + R_{sa} f_{sa}(t) + p_{sa}(t) = 0 \tag{8.29}$$

The flow rates are based on absolute pressures. To compute absolute pressures in intrathoracic compliances, average intrathoracic pressure P_{it} is added to the transmural pressures. If the flow rate is between two intrathoracic compliances, such as the left ventricle and the aorta, P_{it} cancels out. To avoid confusion between pulmonary veins and pulmonic valve, the resistance of the latter is represented by R_{pvpa}, also indicating that, in fact, it represents the lumped resistances of the valve and the pulmonary arteries. The mathematical model is completed by the volume balances:

$$\frac{dv_{lv}(t)}{dt} = f_{lv}(t) - f_{oa}(t) \tag{8.30}$$

$$\frac{dv_{ao}(t)}{dt} = f_{ao}(t) - f_{sa}(t) \tag{8.31}$$

$$\frac{dv_{sa}(t)}{dt} = f_{sa}(t) - f_{st}(t) \tag{8.32}$$

$$\frac{dv_{st}(t)}{dt} = f_{st}(t) - f_{esv}(t) \tag{8.33}$$

$$\frac{dv_{esv}(t)}{dt} = f_{esv}(t) - f_{isv}(t) \tag{8.34}$$

$$\frac{dv_{isv}(t)}{dt} = f_{isv}(t) - f_{ra}(t) \tag{8.35}$$

$$\frac{dv_{ra}(t)}{dt} = f_{ra}(t) - f_{rv}(t) \tag{8.36}$$

$$\frac{dv_{rv}(t)}{dt} = f_{rv}(t) - f_{pa}(t) \tag{8.37}$$

$$\frac{dv_{pa}(t)}{dt} = f_{pa}(t) - f_{pv}(t) \tag{8.38}$$

$$\frac{dv_{pv}(t)}{dt} = f_{pv}(t) - f_{la}(t) \tag{8.39}$$

$$\frac{dv_{la}(t)}{dt} = f_{la}(t) - f_{lv}(t) \tag{8.40}$$

Simplifying assumptions included in this mathematical model involve use of the idealized component descriptions in Table 4.6. The presented mathematical model only applies to compliance volumes above the unstressed volumes.

Numerical values of parameters (Table 8.3) are given in the same number of significant digits as presented in Beneken (1965), or in

Structure	Symbol	Value	Symbol	Value	Symbol	Value	Symbol	Value
Heart chamber (timing)	HR	72	T_{as}	0.105	T_{av}	0.010	T_{vs}	0.327
Intrathoracic space	P_{it}	-4						
Left ventricle	E_{lvmin}	0.09	UV_{lv}	60	R_{mv}	0.003		
	E_{lvmax}	4			R_{av}	0.008		
Aorta	E_{ao}	1.43	UV_{ao}	140				
Systemic arteries	E_{sa}	0.556	UV_{sa}	370	R_{sa}	0.06	L_{sa}	0.0007
Systemic tissues	E_{st}	0.262	UV_{st}	185	R_{sart}	0.8		
					R_{sven}	0.2		
Systemic veins	E_{esv}	0.0169	UV_{esv}	1000	R_{sv}	0.09		
	E_{isv}	0.0182	UV_{isv}	1190				
Right atrium	E_{ramin}	0.05	UV_{ra}	30	R_{ra}	0.003		
	E_{ramax}	0.15						
Right ventricle	E_{rvmin}	0.057	UV_{rv}	40	R_{tv}	0.003		
	E_{rvmax}	0.49			R_{pvpa}	0.003		

Pulmonary arteries	E_{pa}	0.233	UV_{pa}	50				
Pulmonary tissues					R_{pt}	0.11		
Pulmonary veins	E_{pv}	0.045	UV_{pv}	350				
Left atrium	E_{lamin}	0.12	UV_{la}	30	R_{la}	0.003		
	E_{lamax}	0.28						

Here units are: heart rate in min^{-1}, timing heart chambers in s, intrathoracic pressure in mm Hg, elastances in mm Hg/ml, unstressed volumes in ml, resistances in mm Hg ml^{-1} s, and inertia in mm Hg ml^{-1} s^2.

TABLE 8.3 Hemodynamic Model Parameters

three significant digits when computed based on a given parameter, for example, certain elastance values from given compliances.

The sum of R_{sart} and R_{sven} is equal to the resistance of the systemic peripheral vessels as given by Beneken. Their ratio reflects the distribution over arterioles and venules, and is consistent with a functional capillary pressure of approximately 17 mm Hg (Guyton, 2006). The values of E_{st} and UV_{st} were chosen to be consistent with that pressure, and with a total volume in tissue capillaries of 250 ml. The units of parameters listed in Table 8.3 are consistent with the variables: time in seconds, cardiac elastances in millimeter of mercury per milliliter, pressures in millimeter of mercury, flow rates in milliliter per second, and volumes in milliliter.

Considering the input vector from the elastance generator (Fig. 8.2)

$$u(t) = \left[e_{la}(t)\, e_{ra}(t)\, e_{lv}(t)\, e_{rv}(t)\right]^T \tag{8.41}$$

and the state vector

$$x(t) = \left[f_{sa}(t)\, v_{lv}(t)\, v_{ao}(t)\, v_{sa}(t)\, v_{st}(t)\, v_{esv}(t)\, v_{isv}(t)\, v_{ra}(t)\, v_{rv}(t)\, v_{pa}(t)\, v_{pv}(t)\, v_{la}(t)\right]^T \tag{8.42}$$

we observe that Eqs. (8.7) through (8.40) can be put in the form of the (vector) state equation

$$\dot{x}(t) = f(x(t), u(t)) \tag{8.43}$$

To achieve this, the pressure equations (8.7) through (8.17) need to be substituted in the flow rate equations (8.18) through (8.28). Eq. (8.20) is one of the state equations. By substituting the resulting flow rate equations (8.18) and (8.19), and (8.21) through (8.28) into the volume balances (8.30) through (8.40) we find the other eleven state equations. In general, we will not use such a formulation for software implementation; it would reduce the model to only twelve equations, but they would be quite complex. Such a formulation would not be easy to verify or modify, nor would it necessarily lead to more efficient code. The interest of this observation lies in its consequences for the subsequent step in the modeling and simulation process: Software implementation can be based on the algorithms presented in Chap. 5,[5] and only the state variables need to be initialized before the start of a simulation, or saved at the end of a run, to be able to recover the state of the system at the exact same point. Another consequence of this observation is that, using Table 4.1, we see that

[5] An Euler method with integration step size T = 0.001 s gives satisfactory results.

the state equation is nonlinear and time invariant. An example of nonlinearity is the multiplication of the input variable $e_{lv}(t)$ and the state variable $v_{lv}(t)$ in Eq. (8.7). Therefore, multiplying the input by a factor will, in general, not result in multiplication of the state by the same factor. However, a delayed input will result in a delayed—but otherwise unmodified—state.

To complete the model description we would have to give a set of initial values for the state variables. In this particular case, we can simply put all state variables to 0, with the exception of the volume in the extrathoracic systemic veins $v_{esv}(t)$, which we initialize to the total blood volume of 4990 ml. In a few simulated cardiac cycles, the blood volume will distribute across the circulation, and the system will settle into a steady state.

We now consider the general input and output variables as specified in Table 8.1. The repeat of the elastance functions is governed by the heart period $hp(n) = 60/\mathrm{hr}(n)$ in seconds. Beneken provides the following empirical relationships between $hp(n)$ and the systole durations used in Eqs. (8.1) through (8.6):

$$t_{as}(n) = 0.03 + 0.09\,hp(n) \tag{8.44}$$

$$t_{vs}(n) = 0.16 + 0.20\,hp(n) \tag{8.45}$$

$t_{as}(n)$ and $t_{vs}(n)$ are intermediate variables of the elastance generator (Fig. 8.2), and they do not constitute new, independent, input variables. Myocardial contractility is given as the maximum elastance of the left ventricle, and we substitute E_{lvmax} in Eq. (8.5) by

$$e_{lvmax}(n) = mc(n) \tag{8.46}$$

The contractility of the other heart chambers is made to change proportionally by introducing the intermediate variables

$$e_{lamax}(n) = mc(n)E_{lamax} / E_{lvmax} \tag{8.47}$$

$$e_{ramax}(n) = mc(n)E_{ramax} / E_{lvmax} \tag{8.48}$$

$$e_{rvmax}(n) = mc(n)E_{rvmax} / E_{lvmax} \tag{8.49}$$

To take into account fluid administration, $f_{fa}(t)$ becomes an input variable to the basic hemodynamic model, and Eq. (8.34) is updated as follows:

$$\frac{dv_{esv}(t)}{dt} = f_{esv}(t) - f_{isv}(t) + f_{fa}(t)/60.0 \tag{8.50}$$

The factor 60.0 represents the unit conversion from milliliter per minute to milliliter per second, which is used in the model equations. To take into account a time varying intrathoracic pressure, the parameter P_{it} in Eqs. (8.20) and (8.23) is substituted by the input variable 0.760 $p_{it}(t)$. The factor 0.760 represents the conversion from centimeter of water to millimeter of mercury. To take into account the cardiovascular control effector systemic arteriolar resistance, the parameter R_{sart} in Eqs. (8.21) is substituted by the input variable r_{sart}(n), with a constant value over the duration of the cardiac cycle. Variable venous unstressed volume is distributed over the extra- and intrathoracic compartments via the intermediate variables:

$$uv_{esv}(n) = \frac{UV_{esv}}{UV_{esv} + UV_{isv}} v_{vu}(n) \tag{8.51}$$

$$uv_{isv}(n) = \frac{UV_{isv}}{UV_{esv} + UV_{isv}} v_{vu}(n) \tag{8.52}$$

used in Eqs. (8.11) and (8.12), respectively. This completes the adaptation of the model to the input variables listed in Table 8.1. Table 8.4 lists the relationships between basic hemodynamic model variables and output variables of the overall model. Unit conversions or averaging over the cardiac cycle may apply, and all output

Overall model output variable	Basic hemodynamic model variable
$map(n)$	average ($p_{ao}(t)+p_{it}(t)$)
$p_s(t)$	$p_{ao}(t)+p_{it}(t)$
$p_{cv}(t)$	$p_{ra}(t)+p_{it}(t)$
$p_p(t)$	$p_{pa}(t)+p_{it}(t)$
$p_{pcw}(t)$	$p_{pv}(t)+p_{it}(t)$
$co(k)$	average 0.060 $f_{pa}(t)$
$f_{pt}(t)$	$f_{pv}(t)$
$f_{ap}(t)$	$f_{pv}(t)$
$f_{st}(t)$	$f_{st}(t)$
$f_{vp}(t)$	$f_{esv}(t)$
$v_{ap}(t)$	$v_{pv}(t)+v_{la}(t)+v_{lv}(t)+v_{ao}(t)+v_{sa}(t)$
$v_{st}(t)$	$v_{st}(t)$
$v_{vp}(t)$	$v_{esv}(t)+v_{isv}(t)+v_{ra}(t)+v_{rv}(t)+v_{pa}(t)$

TABLE 8.4 Overall Model Output Variables

pressures are absolute rather than transmural pressures. Owing to the absence of a compliance in the pulmonary circulation, the inflow rate of the pulmonary veins and the arterial pool is equal to the inflow rate of the pulmonary tissues. This makes one of these two output variables redundant, but both are maintained to increase readability and facilitate evolution of the model.

For documentation of a mathematical model, we often provide a summary of input, output, and state variables, equations, and numerical values of parameters and initial state.

References and Further Reading

Beneken JEW. A mathematical approach to cardiovascular function. The uncontrolled human system [thesis]. The Netherlands: University of Utrecht; 1965.

Goodwin JA, van Meurs WL, Sá Couto CD, Beneken JEW, Graves SA. A model for educational simulation of infant cardiovascular physiology. *Anest Analg*. Dec 2004; 99(6):1655–1664.

Guyton AC, Hall JE. *Textbook of medical physiology*, 11th ed. Philadelphia: W.B. Saunders; 2006.

Van Meurs WL, Neto P, Azevedo H, Sá Couto CD. "Stan Vintage": A baseline patient for the Human Patient Simulator™ with hemodynamic parameters from the scientific literature [abstract]. *Simul Healthcare*. Fall 2006; 1(3):183.

Zijlmans M, Sá-Couto CD, van Meurs WL, Goodwin JA, Andriessen P. Corrected and improved model for educational simulation of neonatal cardiovascular pathophysiology [technical report]. *Simul Healthc*. Spring 2009; 4(1):49–53.

CHAPTER 9

Respiration

In this chapter we further specify the requirements of the lung mechanics and uptake and distribution models. We introduce a conceptual model consisting of a muscle pressure generator and a model for basic lung mechanics. Lung mechanics are represented via a pneumatic component diagram. We also introduce a conceptual model for uptake and distribution, based on homogenous compartments. The latter takes the form of a multiple model with time-varying fluid volumes and flow rates generated by separate models driving the uptake and distribution model. Complete mathematical models for both subsystems are presented and are shown to have a state variable form.

In this chapter we present models for lung mechanics and uptake and distribution. In the latter model, gas and blood flow rates and volumes may be variable. In Sec. 9.2 we therefore expand the theory presented in Sec. 4.5, introducing the concept of multiple models.

9.1 Model Requirements

From Fig. 7.2 and Table 7.2 we obtain the following overall block diagrams of the lung mechanics and uptake and distribution models (Fig. 9.1).

Tables 9.1 and 9.2 further specify the input and output variables.

We model control of spontaneous breathing via two effector variables: respiratory rate and respiratory muscle pressure amplitude. As for control of circulation, this choice includes the assumption that, for the intended purpose, control of breathing can modeled without a more detailed representation of the autonomic nervous system.[1] Respiratory rate and respiratory muscle pressure have a value once per respiratory cycle. The outputs are continuous-time variables. In Table 7.2 we noted that the overall output variable, tidal volume, is implicit in the gas flow rates and volumes.

With the exception of the partial pressure of inspired carbon dioxide, necessary to simulate re-breathing of CO_2, all input variables

[1] Also note the role of the voluntary nervous system, which has a more direct influence on spontaneous breathing than on the circulation.

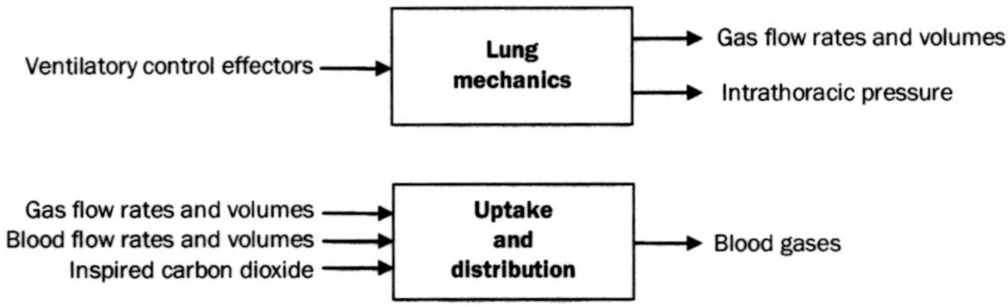

FIGURE 9.1 Overall bock diagrams of the respiratory models.

Input variable	Symbol	Unit	Origin
Respiratory rate	$rr(m)$	min^{-1}	Control of breathing
Muscle pressure amplitude	$mpa(m)$	cm H_2O	Control of breathing
Output variable	**Symbol**	**Unit**	**Destination**
Intrathoracic pressure	$p_{it}(t)$	cm H_2O	Hemodynamics
Alveolar flow rate	$f_A(t)$	ml s^{-1}	Uptake and distribution
Alveolar volume	$v_A(t)$	ml	Uptake and distribution

Here m represents the respiratory cycle.

TABLE 9.1 Input and Output Variables of the Lung Mechanics Model

Input variable	Symbol	Unit	Origin
Partial pressure of inspired carbon dioxide	$p_ICO_2(t)$	mm Hg	Exterior
Alveolar gas flow rate	$f_A(t)$	ml s^{-1}	Lung mechanics
Alveolar gas volume	$v_A(t)$	ml	Lung mechanics
Pulmonary tissues blood flow rate	$f_{pt}(t)$	ml s^{-1}	Hemodynamics
Arterial pool blood flow rate	$f_{ap}(t)$	ml s^{-1}	Hemodynamics
Systemic tissues blood flow rate	$f_{st}(t)$	ml s^{-1}	Hemodynamics
Venous pool blood flow rate	$f_{vp}(t)$	ml s^{-1}	Hemodynamics
Arterial pool blood volume	$v_{ap}(t)$	ml	Hemodynamics
Systemic tissues blood volume	$v_{st}(t)$	ml	Hemodynamics
Venous pool blood volume	$v_{vp}(t)$	ml	Hemodynamics

TABLE 9.2 Input and Output Variables of the Uptake and Distribution Model

Output variable	Symbol	Unit	Destination
Partial pressure of oxygen in arterial blood	$p_aO_2(m)$	mm Hg	Exterior and control of breathing
Partial pressure of carbon dioxide in arterial blood	$p_aCO_2(m)$	mm Hg	Exterior and control of breathing
Partial pressure of oxygen in venous blood	$p_vO_2(m)$	mm Hg	Exterior
Partial pressure of carbon dioxide in venous blood	$p_vCO_2(m)$	mm Hg	Exterior
Oxygen saturation in arterial blood	$s_aO_2(m)$	%	Exterior

Here, depending on the variable and the destination, m represents intermittent blood gas sampling or sampling by the control of breathing model of continuous-time variables.

TABLE 9.2 Input and Output Variables of the Uptake and Distribution Model (*continued*)

were presented previously. For the blood gases, "arterial" and "venous" refer to the systemic circulation. The effects of ambient temperature and relative humidity of inspired gas will be ignored. Other respiratory model requirements, already mentioned in Sec. 7.1, include representation of basic airway, lung, and chest wall mechanics, and pulmonary uptake and distribution of the respiratory gases oxygen and carbon dioxide.

9.2 Multiple Models

In Sec. 4.5 we derived a number of elementary mathematical models for homogeneous compartments. In these models, gas or blood flow rates and volumes were assumed constant. For a number of modeling and simulation purposes, including the one developed here, this may not be a realistic assumption. Consider, for example, the influence of the normal respiratory cycle on alveolar gas composition, or critical incidents such as apnea or significant blood loss. In this chapter we will therefore lift those restrictions. For example, the amount of gas with binding in a perfused blood-tissue compartment with metabolism (Fig. 9.2) now reflects the time-varying blood volume.

$$a(t) = v_b(t)c_b\left[p(t)\right] + V_t c_t\left[p(t)\right] \tag{9.1}$$

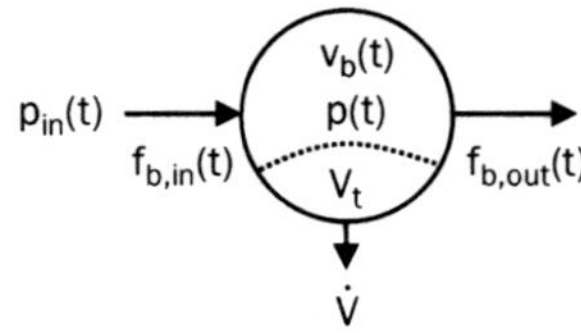

FIGURE 9.2 Homogeneous compartment with time-varying volume and flow rates.

The evolution of this volume depends on the fluid inflow and outflow rates.

$$\frac{dv_b(t)}{dt} = f_{b,in}(t) - f_{b,out}(t) \tag{9.2}$$

The mass balance of the gas also reflects the variable fluid inflow and outflow rates.

$$\frac{da(t)}{dt} = f_{b,in}(t)c_b\left[p_{in}(t)\right] - f_{b,out}(t)c_b\left[p(t)\right] - \dot{V} \tag{9.3}$$

The derivative of the amount now takes into account the time-varying volume.

$$\frac{da(t)}{dt} = \frac{dv_b(t)}{dt}c_b\left[p(t)\right] + v_b(t)\left.\frac{dc_b(p)}{dp}\right|_{p(t)}\frac{dp(t)}{dt} + V_t\left.\frac{dc_t(p)}{dp}\right|_{p(t)}\frac{dp(t)}{dt} \tag{9.4}$$

Substituting Eq. (9.2) in Eq. (9.4) we find:

$$\frac{da(t)}{dt} = \left[f_{b,in}(t) - f_{b,out}(t)\right]c_b\left[p(t)\right] + \left[v_b(t)\left.\frac{dc_b(p)}{dp}\right|_{p(t)} + V_t\left.\frac{dc_t(p)}{dp}\right|_{p(t)}\right]\frac{dp(t)}{dt} \tag{9.5}$$

Equating the right hand sides of Eqs. (9.3) and (9.5), the terms $-f_{b,out}(t)c_b[p(t)]$ cancel out, and we find:

$$\frac{dp(t)}{dt} = \frac{f_{b,in}(t)\left\{c_b\left[p_{in}(t)\right] - c_b\left[p(t)\right]\right\} - \dot{V}}{\left[v_b(t)\left.\frac{dc_b(p)}{dp}\right|_{p(t)} + V_t\left.\frac{dc_t(p)}{dp}\right|_{p(t)}\right]} \tag{9.6}$$

This equation is quite similar to Eq. (4.43), but now reflects the time-varying blood volume and flow rate. Moreover, it specifies that the considered flow rate is the inflow rate; outflow affects the amount, but not the partial pressure. Similar changes apply to the mathematical descriptions of other elementary homogeneous compartments. The time-varying fluid volumes and flow rates may be generated by a separate model, for example, a hemodynamic model (Fig. 9.3).

Beneken and Rideout (1968) presented a generic combined model consisting of coupled hemodynamic and distribution models. This multiple model has matching hemodynamic compliances and homogeneous distribution compartments. Figure 9.4 shows a conceptual multiple model using the symbols introduced in Chap. 4.

For modularity, the mathematical models constituting a multiple model are often kept separate. Note, for example, the similarity between Eq. (9.2) and the volume updates, part of the hemodynamic model described in the previous chapter. If the flow rate is bidirectional, or can be reversed under specific conditions, then the mathematical uptake and distribution model has to be adapted to reflect the changes in supply and evacuation of the transported substance. For example, the mass balance of the first compartment in Fig. 9.4 is adapted as follows, using the unit step function, introduced in Sec. 8.3, as a switch to select the upstream compartment:

$$\begin{aligned}\frac{da_1(t)}{dt} &= f_1(t)\left\{u\left[f_1(t)\right]c_b\left[p_{in}(t)\right]+u\left[-f_1(t)\right]c_b\left[p_1(t)\right]\right\}\\ &\quad - f_2(t)\left\{u\left[f_2(t)\right]c_b\left[p_1(t)\right]+u\left[-f_2(t)\right]c_b\left[p_2(t)\right]\right\}\end{aligned} \tag{9.7}$$

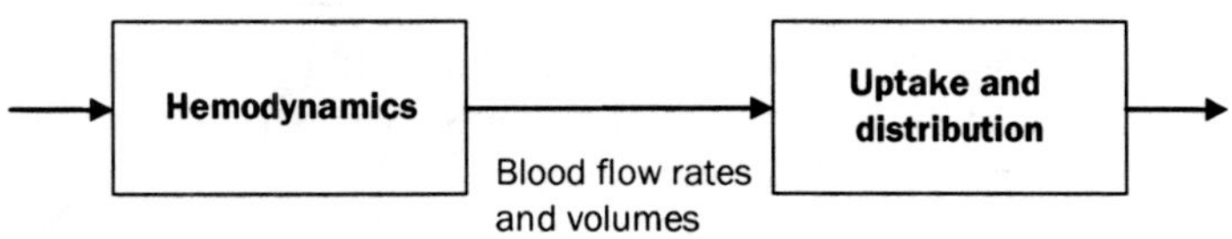

FIGURE 9.3 Block diagram of coupled hemodynamic and uptake and distribution models.

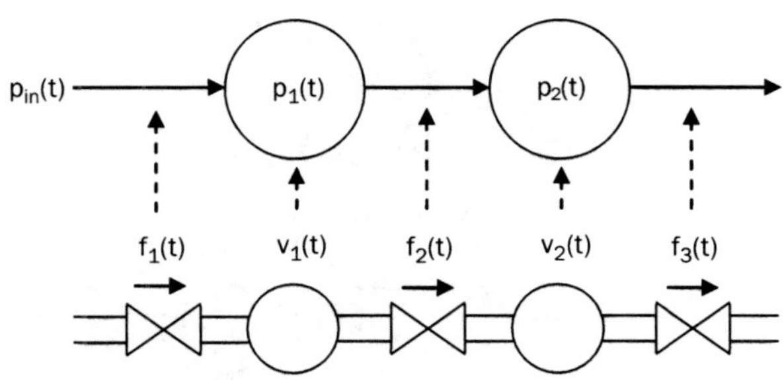

FIGURE 9.4 Basic conceptual multiple model.

As for the derivation of Eq. (9.6), only the terms reflecting the upstream compartments are maintained in the partial pressure equation.

There are several variations on the basic multiple model. In the next section, we will see an example of perfusion with a variable flow rate, but at a fixed volume. In that case, the compliance can be eliminated. The variations in uptake and distribution variables caused by cardiac or respiratory cycle fluctuations in carrier volumes and flow rates may be very small, or not of interest, in a particular modeling and simulation project. In such cases, the flow rates and volumes can be averaged before their use in uptake and distribution computations, resulting in a larger integration step size and more efficient code for these equations. Concentrations or partial pressures in sites corresponding to fluid dynamic compliances may be very similar, or their differences may not be of interest. For example, the partial pressure of oxygen in the left ventricle only has a minute delay with respect to the partial pressure in the left atrium. The volumes of these two compliances can be lumped into a single homogeneous compartment (Fig. 9.5).

The compartment volume is equal to the sum of compliance volumes and the compartment inflow rate corresponds to the inflow rate of the first compliance. This leads to a simplification of the uptake and distribution model and more efficient code. In Sec. 8.1 we observed that if flow rates and blood volumes in specific organs or tissue groups can be considered a fixed fraction of total flow rate and volume, then they can be computed as such. In a multiple model this results in a single hemodynamic element (resistance or compliance) corresponding to several distribution compartments. If parallel organs or tissue groups have different hemodynamic characteristics, or if incorporation of local control mechanisms depending on compartment partial pressures is required, such a scheme cannot be used, and parallel hemodynamic and distribution elements are required, for example (Fig. 9.6).

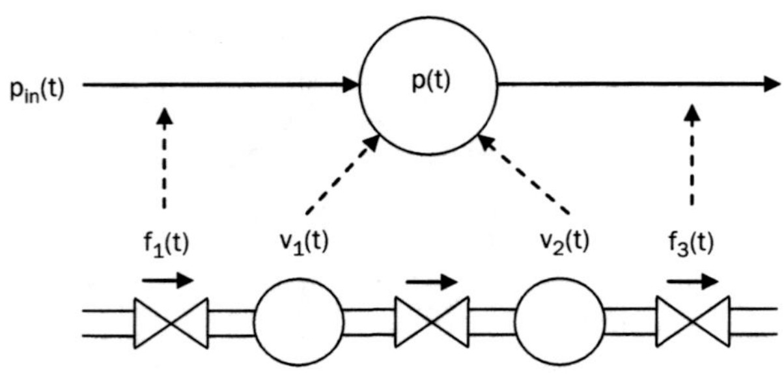

FIGURE 9.5 Multiple model with lumped compartments.

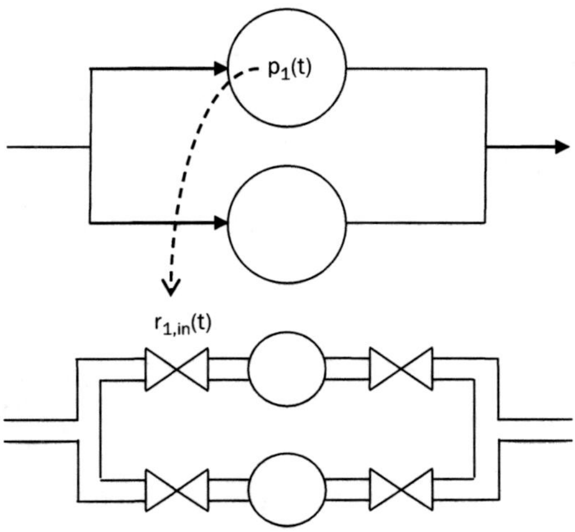

FIGURE 9.6 Multiple model with parallel tissue groups and autoregulation in one.

An example of such autoregulation would be the coronary circulation where, according to one hypothesis, a decrease in oxygen concentration leads to the release of a vasodilator substance affecting the arterioles (Guyton, 2006; also see Fig. 9.6). This also constitutes an example of feedback from the distribution model to the carrier model. A multiple model may involve several carriers (the fluids, gas or blood), even for a single compartment, for example, the elementary ventilated and perfused compartment in Table 4.8. It may also involve several substances (a particular gas, drug, hormone, heat) using the same underlying carrier model(s) (Fig. 9.7).

9.3 Conceptual Models

Mecklenburgh et al. (1992) proposed a lung model with direct representation of respiratory muscle activity. Similar to hemodynamics (Sec. 8.2) the electromechanical processes that are responsible for the generation of the time-varying respiratory muscle pressure $p_{MUS}(t)$ can be separated from the pneumatic aspects of lung mechanics (Fig. 9.8).

In the next section we will present an empirical muscle pressure generator that uses the input variables respiratory rate and muscle pressure amplitude, available once per respiratory cycle, to parameterize the continuous-time respiratory muscle pressure waveform presented by Mecklenburgh et al. (1992). Their model for basic lung mechanics consists of a resistance in series with a compliance. The resistance R reflects resistance to flow in the airways, as well as energy losses in moving lung and chest wall tissues. As

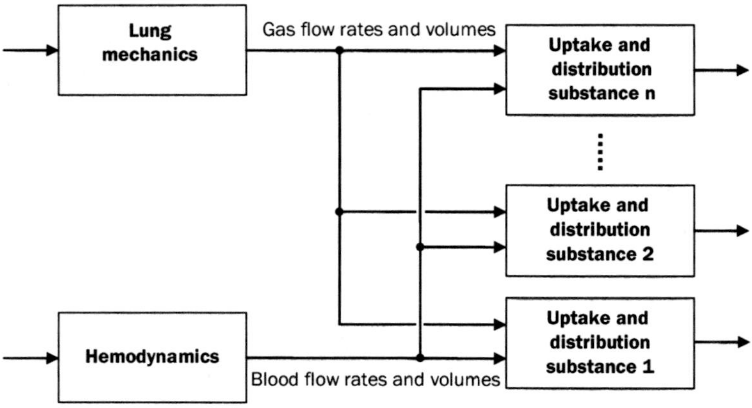

FIGURE 9.7 Block diagram of a multiple model with two carriers and *n* substances.

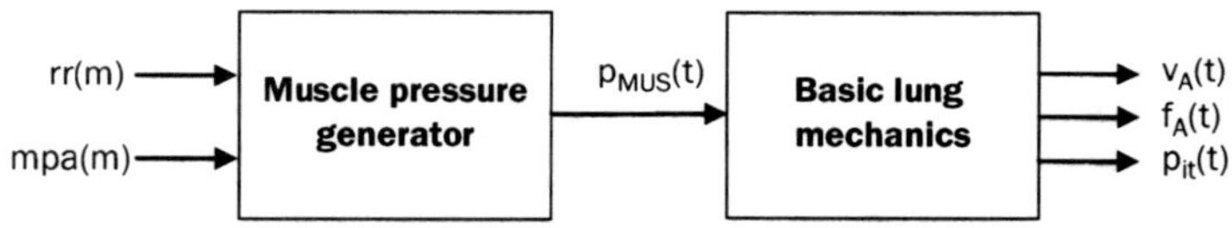

FIGURE 9.8 Block diagram of the lung mechanics model.

simulating the intrathoracic pressure is part of the requirements, we expand this model by dividing the compliance into lung compliance C_L and chest wall compliance C_{CW}. Figure 9.9 shows the corresponding component diagram, already in the form of a multiple model including gas mixing in the airways.

During spontaneous inspiration, respiratory muscle pressure contributes to the transmural pressure of the chest wall compliance, resulting in a negative intrapleural pressure, which we will consider to be equal to the intrathoracic pressure. This negative pressure contributes to the transmural pressure of the lung compliance, resulting in a negative alveolar pressure and gas flow across the airway resistance. On expiration, the muscle pressure is removed, and the volume accumulated in the compliances creates a positive alveolar pressure and passive exhalation.

The pulmonary uptake and distribution model consists of two homogeneous compartments coupled to this model: alveolar space, where gas exchange $e(t)$ takes place, and dead space, without gas exchange. Ventilation is bidirectional; the arrow above the resistance defines positive flow rate. The dead space is considered to have a fixed volume V_D, hence the absence of a corresponding compliance in the lung mechanics model. The alveolar space has a time-varying

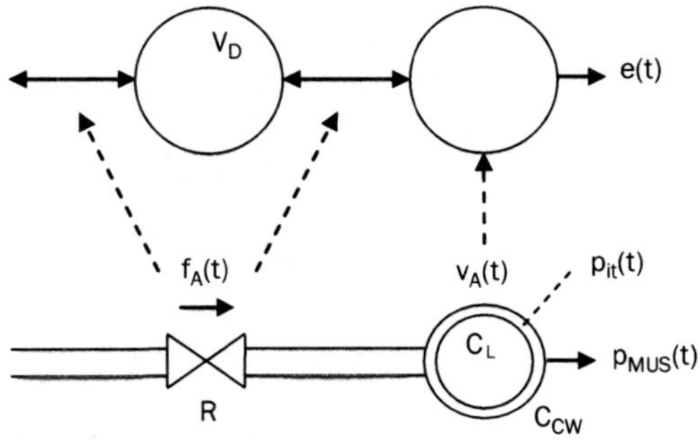

Figure 9.9 Multiple model of lung ventilation and gas uptake.

volume corresponding to the volume of the compliance. Gas exchange is considered variable, depending on pulmonary perfusion, venous blood gas content, and alveolar partial pressure. The partial pressures in blood leaving the alveolar capillaries are modified by admixture of blood coming directly from the pulmonary arteries (Fig. 9.10).

Normally this admixture is a small fraction of total flow rate, in the order of 2 percent, but in certain critical incidents or pathologies it can increase significantly. The shunt flow rate and the flow rate through the compartment are equal to $SF\ f_{pt}(t)$ and $(1 - SF)\ f_{pt}(t)$, respectively.

The concept of lumping hemodynamic compliances into a single distribution compartment is presented in Fig. 9.5. For modeling distribution of respiratory gases, hemodynamic compliances can be lumped as listed in Table 8.2. Figure 9.11 shows these compartments in association with the compliances of the hemodynamic model (Fig. 8.3) and of the model of lung ventilation (Fig. 9.9).

Figure 9.12 shows the resulting uptake and distribution model, using the homogeneous compartments introduced in Sec. 4.5 and adding the pulmonary shunt of Fig. 9.10.

A simplifying assumption included in this model is that the amount of gas dissolved in pulmonary blood and lung tissue can be ignored with respect to the amount of gas in the alveolar gas mixture; the homogenous compartment has no tissues, and the separation between gas and blood included in the model in Table 4.8 is removed. For the same reason, there is no compliance corresponding to a time-varying pulmonary blood volume (Figs. 9.10 and 9.11). The main simplifying assumption included in the lung ventilation and gas uptake model (Fig. 9.9) is that such a simple model can indeed represent the essential aspects of spontaneous breathing and gas mixing in the airways. The absence of representation of separate left and right lungs is consistent with the requirements, which do not include simulation of unilateral incidents, such as pneumothorax.

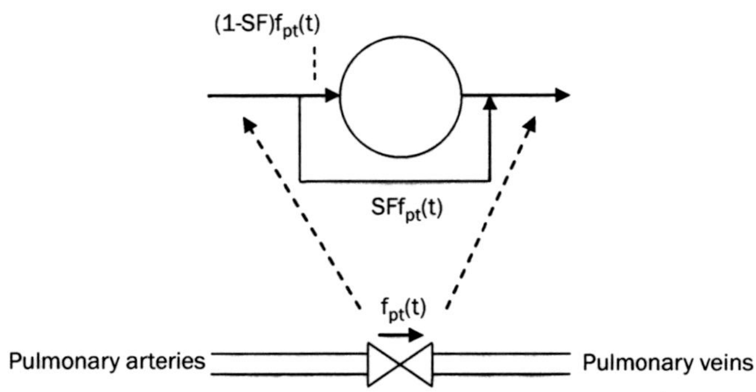

FIGURE 9.10 Multiple model of lung perfusion including a simple shunt with shunt fraction *SF*, $0 \leq SF \leq 1$.

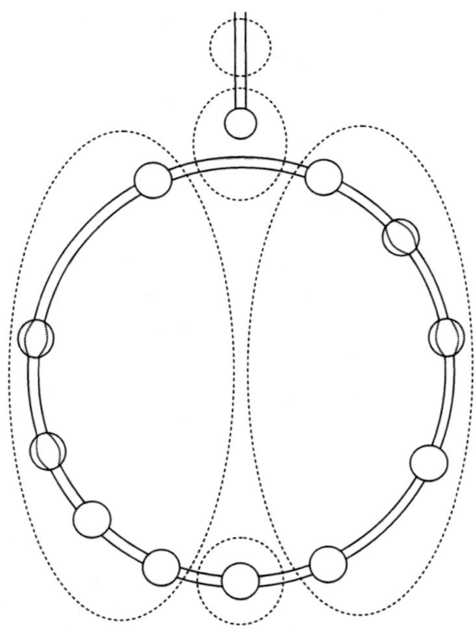

FIGURE 9.11 Multiple model of ventilation, circulation, and uptake and distribution of respiratory gases.

The effect of gas warming and humidification during inspiration on alveolar volume is ignored, but could be incorporated using the concepts presented in Sec. 4.5. Simulation of mechanical ventilation is not part of the requirements, but would result in an additional input to the basic lung mechanics model in Fig. 9.8.

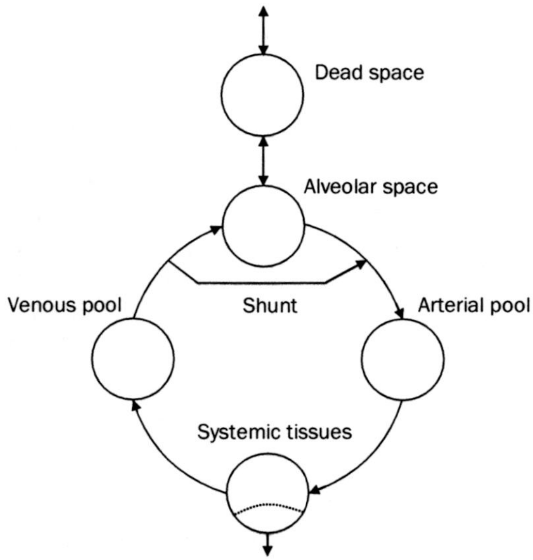

FIGURE 9.12 Uptake and distribution model.

9.4 Mathematical Models

The muscle pressure curve presented by Mecklenburgh et al. (1992) can be parameterized empirically as a piece-wise continuous function consisting of a ramp and a decaying exponential with amplitude *mpa*(*m*), Eq. (9.8).

$$p_{MUS}(t) = \begin{cases} mpa(m)\dfrac{t}{t_I(m)}, & 0 \le t < t_I(m) \\ mpa(m)\dfrac{e^{-\frac{t-t_I(m)}{4t_E(m)}} - e^{-\frac{1}{4}}}{1-e^{-\frac{1}{4}}}, & t_I(m) \le t < t_I(m) + t_E(m) \end{cases} \tag{9.8}$$

The inspiratory and expiratory times are defined with respect to the muscle pressure waveform, rather than with respect to flow rate. They are considered fixed fractions of the respiratory period in seconds: $60/rr(m)$, Eqs. (9.9) and (9.10).

$$t_I(m) = \frac{1}{3}\frac{60}{rr(m)} \tag{9.9}$$

$$t_E(m) = \frac{2}{3}\frac{60}{rr(m)} \tag{9.10}$$

It can be verified that this waveform is continuous in the knots $t = 0$, $t = t_I$, and $t = t_I + t_E = RP$, where t is the time in the respiratory cycle and RP the respiratory period. The muscle pressure waveform is repeated with RP. The basic lung mechanics model takes this waveform as an input (Fig. 9.8). Following the same procedure as for the hemodynamic model (Sec. 8.3), we derive a mathematical model corresponding to the conceptual model of Fig. 9.9. Considering the state variable $v_A(t)$ and the functional residual capacity (FRC), the transmural pressure of the lung compliance is:

$$p_L(t) = -P_{ITE} + \frac{v_A(t) - FRC}{C_L} \tag{9.11}$$

This represents a minor adaptation of the mathematical description of the compliance listed in Table 4.6. The nonlinearities of this relationship are not considered here. The parameter P_{ITE} indicates the intrathoracic or intrapleural pressure at the end of a passive expiration with $v_A(t) = FRC$. At that point, the alveolar pressure is equal to atmospheric pressure, which we will take as our reference pressure, and the lungs are kept open by a positive transmural pressure, due to a slightly negative pressure in the intrapleural space. Assuming that the intrapleural volume can be ignored, the transmural pressure of the chest wall compliance is:

$$p_{CW}(t) = P_{ITE} + \frac{v_A(t) - FRC}{C_{CW}} \tag{9.12}$$

At the end of a passive expiration the chest wall has a slightly negative transmural pressure, and is pulled inward due to the pressure in the intrapleural space. Compressibility of respiratory gases is included in these compliances. The pressure in the alveolar space is the sum of the muscle pressure and the two transmural pressures [respecting the sign of $p_{MUS}(t)$][2]:

$$p_A(t) = p_L(t) + p_{CW}(t) - p_{MUS}(t) \tag{9.13}$$

Flow rate through the linear resistance is equal to:

$$f_A(t) = -\frac{p_A(t)}{R} = \frac{-p_L(t) - p_{CW}(t) + p_{MUS}(t)}{R} \tag{9.14}$$

[2] A form of Kirchhoff's voltage law.

Substituting the transmural pressures

$$f_A(t) = -\frac{v_A(t) - FRC}{RC_L} - \frac{v_A(t) - FRC}{RC_{CW}} + \frac{p_{MUS}(t)}{R} \tag{9.15}$$

and reordering terms, the flow rate can be written in terms of the state and input variables:

$$f_A(t) = -\frac{C_L + C_{CW}}{C_L C_{CW} R}[v_A(t) - FRC] + \frac{1}{R} p_{MUS}(t) \tag{9.16}$$

Note that this equation includes the formula for in series compliances, Table 4.14. The volume balance is simply:

$$\frac{dv_A(t)}{dt} = f_A(t) \tag{9.17}$$

Per the requirements, this equation does not contain compensations for differences between ambient and body temperature, or for differences in humidity between inspired and expired gases. Substituting the flow rate we find:

$$\frac{dv_A(t)}{dt} = -\frac{C_L + C_{CW}}{C_L C_{CW} R}[v_A(t) - FRC] + \frac{1}{R} p_{MUS}(t) \tag{9.18}$$

The state variable $v_A(t)$ is one of the output variables. The flow rate $f_A(t)$, given by Eq. (9.16), is another. The output variable $p_{it}(t)$ is the sum of the muscle pressure and the transmural pressure of the chest wall.

$$p_{it}(t) = p_{CW}(t) - p_{MUS}(t) \tag{9.19}$$

Substituting the transmural pressure, we find:

$$p_{it}(t) = P_{ITE} + \frac{v_A(t) - FRC}{C_{CW}} - p_{MUS}(t) \tag{9.20}$$

Considering the input $u(t) = p_{MUS}(t)$, the state $x(t) = v_A(t) - FRC$, and the output vector $y(t) = [v_A(t) - \text{FRC} \;\; f_A(t) \;\; p_{it}(t) - P_{ITE}]^T$, we observe that:

$$\frac{dx(t)}{dt} = \frac{dv_A(t)}{dt} \tag{9.21}$$

and Eq. (9.18) is a state equation of the form

$$\frac{dx(t)}{dt} = ax(x) + bu(t) \tag{9.22}$$

with scalar constants a and b that can be read from the equation. The transmission of the state variable to the output, and Eqs. (9.16) and (9.20) can be written as

$$\begin{pmatrix} v_A(t) - FRC \\ f_A(t) \\ p_{it}(t) - P_{ITE} \end{pmatrix} = \begin{pmatrix} 1 \\ -\dfrac{C_L + C_{CW}}{C_L C_{CW} R} \\ \dfrac{1}{C_{CW}} \end{pmatrix} [v_A(t) - FRC] + \begin{pmatrix} 0 \\ \dfrac{1}{R} \\ -1 \end{pmatrix} p_{MUS}(t) \tag{9.23}$$

which is an output equation of the form

$$y(t) = cx(x) + du(t) \tag{9.24}$$

with vector constants c and d that can be read from the equation. Numerical values of model parameters are given in Table 9.3, in a consistent set of units.

Name	Symbol	Value	Unit
Respiratory muscle pressure amplitude	MPA	4.5	cm H_2O
Respiratory period	RP	5	s
Airways and lung tissue resistance	R	3.7 10^{-3}	cm H_2O ml^{-1} s
Functional residual capacity	FRC	2300	ml
Lung compliance	C_L	200	ml/cm H_2O
End-expiratory pressure in the intrapleural space	P_{ITE}	−4	cm H_2O
Chest wall compliance	C_{CW}	244	ml/cm H_2O

Table 9.3 Lung Mechanics Model Parameters

Mecklenburgh et al. (1992) present numerical values of model parameters for only three individual subjects. Guyton and Hall (2006) give numerical values for most parameters, reflecting a normal adult male. The value of C_{CW} is derived from a total lung and chest wall compliance of 110 ml/cm H_2O and the value of C_L. The timing and amplitude of the muscle pressure waveform are governed by the input variables *mpa*(*n*) and *rr*(*n*). Table 9.3 lists corresponding baseline values for the muscle pressure amplitude *MPA* and the respiratory period *RP*. For the given set of parameters the value of *MPA* results in a steady-state volume increase of 500 ml, which is the typical value for tidal volume given by Guyton and Hall. The value of *RP* corresponds to a typical respiratory rate of 12 min^{-1}. Guyton and Hall provide two values for P_{ITE}, $P_{ITE} = -4$ cm H_2O is retained here. This value and a negative deflection of approximately 2 cm H_2O, caused by *MPA* divided over the two in series compliances, are consistent with the hemodynamic parameter $P_{it} = -4$ mm Hg ≈ -5 cm H_2O, which indicates the average value of the intrathoracic pressure. Patterson et al. (1968) give a value for the resistance *R* reflecting airways and lung tissues. Resistive losses in the chest wall are therefore ignored. This value is an average value for patients undergoing open heart surgery, with access to the intrapleural space, but not for a typical healthy adult male. A few parameters are included as numerical values in equations (9.8) through (9.10). For some applications, it could be useful to represent those via symbols and list them in a table, similar to Table 9.3. The end-expiratory situation is a convenient initial state: $v_A(0) = FRC$.

The conceptual model of uptake and distribution is given in Fig. 9.12. The mathematical model is derived based on the descriptions of the homogeneous compartments in Table 4.8, and their time-varying equivalents, derived in Sec. 9.2. Considering the bidirectional flow rate $f_A(t)$ (Fig. 9.9), we observe that $u[f_A(t)] = 1$ on inspiration and $u[f_A(t)] = 0$ on expiration. Conversely, $u[-f_A(t)] = 0$ on inspiration and $u[-f_A(t)] = 1$ on expiration. This is used to switch upstream compartments, as in Eq. (9.7). The description of the ventilated compartment in Table 4.8 can be adapted for the dead space, with a fixed volume but a variable flow rate.

$$\frac{dp_D(t)}{dt} = \frac{u[f_A(t)]f_A(t)p_{in}(t) - u[-f_A(t)]f_A(t)p_A(t) - |f_A(t)|p_D(t)}{V_D} \tag{9.25}$$

This, and all subsequent, equations have specific instantiations for the respiratory gases oxygen O_2 and carbon dioxide CO_2. For oxygen, when breathing room air, the partial pressure of inspired gas is constant: $p_{in}(t) = P_IO_2$, for carbon dioxide $p_{in}(t) = p_ICO_2(t)$. The amount of gas in the alveolar space is equal to:

$$a_A(t) = v_A(t)\frac{p_A(t)}{P_T} \tag{9.26}$$

where P_T represents the total pressure, ignoring the respiratory cycle variations in this quantity. The derivative of the amount is:

$$\begin{aligned}\frac{da_A(t)}{dt} &= \frac{d}{dt}\left\{v_A(t)\frac{p_A(t)}{P_T}\right\} = \frac{dv_A(t)}{dt}\frac{p_A(t)}{P_T} + \frac{v_A(t)}{P_T}\frac{dp_A(t)}{dt}\\ &= f_A(t)\frac{p_A(t)}{P_T} + \frac{v_A(t)}{P_T}\frac{dp_A(t)}{dt}\end{aligned} \tag{9.27}$$

The mass balance reflects ventilatory supply and evacuation of gas, from and to the dead space, and perfusion coming from the venous pool with a flow rate $(1 - SF)f_{pt}(t)$ (also see Fig. 9.10).

$$\begin{aligned}\frac{da_A(t)}{dt} &= f_A(t)\left\{u\left[f_A(t)\right]\frac{p_D(t)}{P_T} + u\left[-f_A(t)\right]\frac{p_A(t)}{P_T}\right\}\\ &+ (1-SF)f_{pt}(t)\left\{c_b\left[p_{vp}(t)\right] - c_b\left[p_A(t)\right]\right\}\end{aligned} \tag{9.28}$$

An assumption included in the last term of this equation is that there is perfect equilibration of partial pressures between the alveolar space and pulmonary venous blood. The relationships between partial pressure and blood gas content for oxygen and carbon dioxide will be discussed as part of the model parameters. Using:

$$f_A(t) = \left\{u\left[f_A(t)\right] + u\left[-f_A(t)\right]\right\}f_A(t) \tag{9.29}$$

in Eq. (9.27) and combining Eqs. (9.27) and (9.28), the expiratory terms cancel out, and we find:

$$\frac{dp_A(t)}{dt} = \frac{f_A(t)u\left[f_A(t)\right]\left\{p_D(t) - p_A(t)\right\} + P_T(1-SF)f_{pt}(t)\left\{c_b\left[p_{vp}(t)\right] - c_b\left[p_A(t)\right]\right\}}{v_A(t)} \tag{9.30}$$

The confluent, Table 4.10, in the pulmonary veins combines the mass flow rates from the alveolar space and the venous pool.

$$c_{pv}(t) = \frac{(1-SF)f_{pt}(t)c_b\left[p_A(t)\right] + SFf_{pt}(t)c_b\left[p_{vp}(t)\right]}{f_{pt}(t)} \tag{9.31}$$

The dynamic equations for the partial pressures in the arterial pool, the systemic tissues, and the venous pool result from adaptations of Eq. (9.6)

$$\frac{dp_{ap}(t)}{dt} = \frac{f_{ap}(t)\left\{c_{pv}(t) - c_b\left[p_{ap}(t)\right]\right\}}{v_{ap}(t)\left.\frac{dc_b(p)}{dp}\right|_{p_{ap}(t)}} \tag{9.32}$$

$$\frac{dp_{st}(t)}{dt} = \frac{f_{st}(t)\left\{c_b\left[p_{ap}(t)\right] - c_b\left[p_{st}(t)\right]\right\} - \dot{V}}{\left[v_{st}(t)\left.\frac{dc_b(p)}{dp}\right|_{p_{st}(t)} + V_t\left.\frac{dc_t(p)}{dp}\right|_{p_{st}(t)}\right]} \tag{9.33}$$

$$\frac{dp_{vp}(t)}{dt} = \frac{f_{vp}(t)\left\{c_b\left[p_{st}(t)\right] - c_b\left[p_{vp}(t)\right]\right\}}{v_{vp}(t)\left.\frac{dc_b(p)}{dp}\right|_{p_{vp}(t)}} \tag{9.34}$$

Only the systemic tissues compartment contains tissue volume and has metabolism. For oxygen, metabolism is positive, indicating consumption, for carbon dioxide it is negative, indicating production.

Considering the input vector

$$u(t) = \left[p_I CO_2(t)\; f_A(t)\; v_A(t)\; f_{pt}(t)\; f_{ap}(t)\; f_{st}(t)\; f_{vp}(t)\; v_{ap}(t)\; v_{st}(t)\; v_{vp}(t)\right]^T \tag{9.35}$$

and the state vector

$$x(t) = \left[p_D CO_2(t)\; p_A CO_2(t)\; p_{ap} CO_2(t)\; p_{st} CO_2(t)\; p_{vp} CO_2(t)\right]^T \tag{9.36}$$

equations (9.25), (9.30) through (9.34) constitute a state equation of the form

$$\dot{x}(t) = f(x(t), u(t)) \tag{9.37}$$

For oxygen, the input $p_I CO_2(t)$ is replaced by the parameter $P_I O_2$. The partial pressure output variables correspond to state variables. For carbon dioxide:

$$\begin{pmatrix} p_a CO_2(t) \\ p_v CO_2(t) \end{pmatrix} = \begin{pmatrix} 0 & 0 & 1 & 0 & 0 \\ 0 & 0 & 0 & 0 & 1 \end{pmatrix} x(t) \tag{9.38}$$

which is an output equation of the form

$$y(t) = Ax(t) \tag{9.39}$$

The output variable oxygen saturation in arterial blood $s_aO_2(t)$ is a nonlinear function, presented below, of the state variable partial pressure of oxygen in arterial blood. Discrete-time versions of these output variables (Table 9.2) are obtained via sampling of the continuous-time equivalents. Table 9.4 lists the uptake and distributions parameters that appear in the state equations.

The first two parameters indicate conditions of the environment, the latter four are patient specific. Volume of perfused tissues is estimated for a 70 kg subject. Oxygen consumption and carbon dioxide production are considered at BTPS, and are related via a respiratory quotient RQ = 0.8. Values for these parameters can be found in Guyton and Hall (2006), or other physiology textbooks. Additional parameters are included in the functions that relate content to partial pressure. The content of oxygen in blood as a function of its partial pressure $c_bO_2[pO_2]$ is computed as follows. Severinghaus (1979) gives an accurate approximation of the relationship between pO_2 in mm Hg and the dimensionless fraction of hemoglobin saturated with oxygen S.

$$S = \left(23400 \left(pO_2^{\,3} + 150\, pO_2 \right)^{-1} + 1 \right)^{-1} \tag{9.40}$$

Oxygen content results from a combination of oxygen bound to hemoglobin and dissolved oxygen. The content in milliliter of oxygen

Name	Symbol	Value	Unit
Atmospheric pressure	P_T	760	mm Hg
Partial pressure of oxygen inspired air	P_IO_2	160	mm Hg
Shunt fraction	SF	0.02	-
Volume of perfused tissues	V_t	60000	ml
Oxygen metabolism	$\dot{V}O_2$	250	ml min^{-1}
Carbon dioxide production	$\dot{V}CO_2$	200	ml min^{-1}

Table 9.4 Uptake and Distribution Model Parameters

at STP per milliliter of blood (see Sec. 4.5) is given by Severinghaus (1979):

$$[O_2]_b = 0.0134\, Hb\, S + 0.000031\, pO_2 \tag{9.41}$$

where *Hb* is the hemoglobin concentration in grams per deciliter. Typical hemoglobin concentrations for males and females are 16 and 14 g/dl, respectively (Guyton, 2006). As we have seen in Sec. 4.5, to obtain content in milliliter of oxygen at BTPS per milliliter of blood, this result needs to be multiplied by a factor 1.21.

$$c_b O_2 = 1.21 [O_2]_b \tag{9.42}$$

The amount of gas in the alveolar space, Eq. (9.26), is at BTPS conditions. The alveolar mass balance Eq. (9.28) and resulting state equation (9.30) also assume use of BTPS content units. The output variable oxygen saturation in arterial blood $s_a O_2(n)$ in percentage can be derived from the continuous-time partial pressure of oxygen in arterial blood, and Eq. (9.40).

$$s_a O_2(t) = 100 \left(23400 \left(p_a O_2(t)^3 + 150 p_a O_2(t) \right)^{-1} + 1 \right)^{-1} \tag{9.43}$$

In the tissues, outside of the circulation, there is no hemoglobin, and considering Eq. (9.33), the amount of dissolved oxygen can be ignored.

$$c_t O_2 = 0 \tag{9.44}$$

Carbon dioxide is transported by the blood in many forms: the dissolved state (7%), as bicarbonate ions HCO_3^- (70%), and bound to hemoglobin (23%). Kelman (1967) gives a procedure to compute the content of carbon dioxide in blood as a function of its partial pressure $c_b CO_2[pCO_2]$. For our application, this procedure can be approximated by a linear relationship, which gives the content in milliliter of carbon dioxide at STP per milliliter of blood.

$$[CO_2]_b = 0.0066\, pCO_2 + 0.23 \tag{9.45}$$

Content in milliliter of carbon dioxide at BTPS per milliliter of blood is given by:

$$c_b CO_2 = 1.21 [CO_2]_b \tag{9.46}$$

The tissues do not contain hemoglobin, but carbon dioxide is present in the other forms. At the same partial pressure:

$$c_t\mathrm{CO}_2 = 0.77\ c_b\mathrm{CO}_2 \tag{9.47}$$

The derivatives of the partial pressure-content relationships can be computed analytically, or empirically using small changes in partial pressure, for example, 1 mm Hg. Dependency of these relationships on other blood gases (Bohr and Haldane effects) is not considered here, but more information can be found in the references and further reading. A more complete analysis of pulmonary gas uptake could also take into account water vapor and nitrogen (N_2). In that case, the sum of partial pressures has to add up to the total pressure in the lung mechanics model, and net uptake has to be taken into account by that model.

References and Further Reading

Barton SA, Black AM, Hahn CE.: Dynamic models and solutions for evaluating ventilation, perfusion, and mass transfer in the lung - Part I: The models. *IEEE Trans Biomed Eng.* Jun 1988; 35(6):450–7.

Beneken JE, Rideout VC. The use of multiple models in cardiovascular system studies: transport and perturbation methods. *IEEE Trans Biomed Eng.* Oct 1968; 15(4):281–9.

Chiari L, Avanzolini G, Ursino M. A comprehensive simulator of the human respiratory system: validation with experimental and simulated data. *Ann Biomed Eng.* Nov-Dec 1997; 25(6):985–99.

Fincham WF, Tehrani FT. A mathematical model of the human respiratory system. *J Biomed Eng.* Apr 1983; 5(2):125–33.

Grodins FS, Buell J, Bart AJ. Mathematical analysis and digital simulation of the respiratory control system. *J Appl Physiol.* Feb 1967; 22(2):260–76.

Guyton AC, Hall JE. *Textbook of medical physiology,* 11th ed. Philadelphia: W.B. Saunders; 2006.

Kellman GR. Digital computer subroutine for the conversion of PCO2 into blood CO2 content. *Respiration Physiol.* 1967; 3:111–15.

Mapleson WW. An electric analogue for uptake and exchange of inert gases and other agents. *Appl Physiol.* Jan 1963; 18:197–204.

Mecklenburgh JS, Al-Obaidi TAA, Mapleson WW. A model lung with direct representation of respiratory muscle activity. *Br. J. Anaesth.* 1992; 68:603–12.

Patterson RW, Sullivan SF. Interrelation between Paco2 and PAco2 in the control of the mechanics of breathing in man. *Bull N Y Acad Med.* Oct 1968; 44(10):1265–72.

Riley RL, Cournand A. 'Ideal' alveolar air and the analysis of ventilation-perfusion relationships in the lungs. *J Appl Physiol.* Jul 1949; 1(12):825–47.

Severinghaus JW. Simple, accurate equations for human blood O2 dissociation computations. *J. Appl. Physiol.* 1979; 46(3):599–602.

CHAPTER 10

Physiologic Control

We consider that an operating point in physiological feedback results from the equilibrium between a system and its control. The models for control of circulation and spontaneous breathing, presented in this chapter, are based on linearization of the relationship between the controlled variable (e.g., mean arterial pressure) and the control effector (e.g., heart rate) in the operating point. The static models consist of piecewise linear receptor functions, control gains, and saturations on the control effectors. Complete mathematical models for both control systems are presented.

One of the benefits of feedback control is that it allows for maintaining variables within narrow ranges, despite changes in system characteristics or environmental conditions. Thus, feedback is at the center of many homeostatic mechanisms, including control of blood pressure and control of blood gases. We consider that an *operating point* in physiological feedback results from the equilibrium between a system and its control. This situation is unlike manmade control systems with an explicit setpoint for the controlled variable (see for example van de Vegte, 1994). Figure 10.1 gives an example in cardiovascular control.

A rise in heart rate causes a rise in blood pressure in the uncontrolled cardiovascular system. In the baroreflex control system, a drop in blood pressure causes a rise in heart rate. Coupling the two systems results in the operating point as indicated in Fig. 10.1. Heart rate acts as a control effector. We will call the value of a control effector in the operating point a reference (REF). Mean arterial blood pressure is the controlled variable. We will also call the value in the operating point a setpoint (SP), despite the fact that it results from the equilibrium between the two systems. A perturbation, for example, acute blood loss, will cause the characteristics of the uncontrolled system to change, as illustrated in Fig. 10.1. Without feedback, heart rate would remain equal to the REF value, and a significant blood pressure drop would result.[1] With feedback, a new equilibrium

[1] A precise analysis would have to take into account the sympathovagal balance in the operating point.

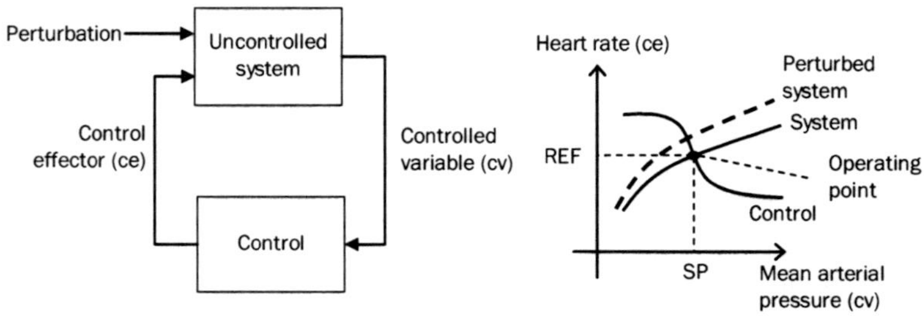

FIGURE 10.1 Simplified block diagram of a physiologic control system and steady-state relationships in baroreflex control.

between the perturbed system and its control is established, resulting in a much smaller decrease in blood pressure. A more intuitive way of looking at this feedback is that, in the coupled systems, a drop in blood pressure results in a rise in heart rate, partially counteracting the original blood pressure drop. In the subsequent sections we will introduce simple empirical models of control of circulation and control of spontaneous breathing.

10.1 Model Requirements

From Fig. 7.2 we obtain the overall block diagrams of the control models as shown in Fig. 10.2.

Table 10.1 defines the input and output variables of the control of circulation model, explicitly listing the control effectors, including the overall output heart rate (Table 7.2). These variables are a subset of the variables of the uncontrolled model, listed in Table 8.1, switching input and output variables.

These are essentially the same input and output variables as for the model presented by Wesseling and Settels (1985). A minor difference is that the directly measurable control effector total peripheral resistance (TPR) is replaced by systemic arteriolar resistance r_{sart}, which is the physiological variable in the Beneken model (Chap. 8), causing a TPR change. A simplifying assumption included in this input-output configuration is that control of circulation only depends on systemic mean arterial blood pressure.

Table 10.2 defines the input and output variables of the control of breathing model, explicitly listing the control effectors, including the overall output respiratory rate (Table 7.2). The input variables are a subset of the output variables of the uptake and distribution model (Table 9.2). The output variables are the input variables of the lung mechanics model (Table 9.1).

Simplifying assumptions included in this input–output configuration are that the control of spontaneous breathing only

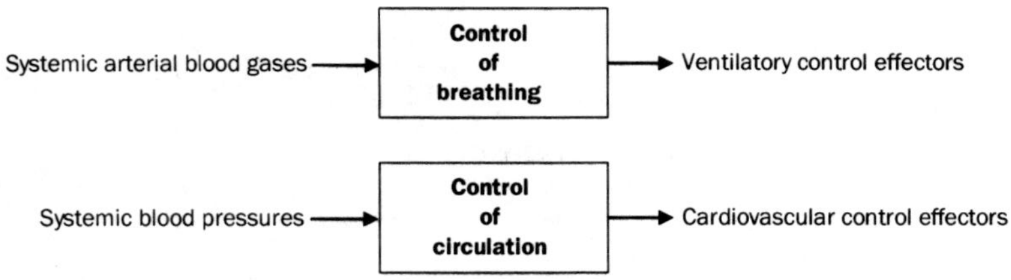

FIGURE 10.2 Block diagrams of physiologic control systems.

Input variable	Symbol	Unit	Origin
Mean arterial blood pressure	*map*(*n*)	mm Hg	Hemodynamics
Output variable	**Symbol**	**Unit**	**Destination**
Heart rate	*hr*(*n*)	min^{-1}	Hemodynamics
Myocardial contractility	*mc*(*n*)	mm Hg ml^{-1}	Hemodynamics
Systemic arteriolar resistance	$r_{sart}(n)$	mm Hg ml^{-1} min	Hemodynamics
Venous unstressed volume	$v_{vu}(n)$	ml	Hemodynamics

Here *n* represents the cardiac cycle.

TABLE 10.1 Input and Output Variables of the Control of Circulation Model

Input variable	Symbol	Unit	Origin
Partial pressure of oxygen in arterial blood	$p_aO_2(m)$	mm Hg	Uptake and distribution
Partial pressure of carbon dioxide in arterial blood	$p_aCO_2(m)$	mm Hg	Uptake and distribution
Output variable	**Symbol**	**Unit**	**Destination**
Respiratory rate	*rr*(*m*)	min^{-1}	Lung mechanics
Muscle pressure amplitude	*mpa*(*m*)	cm H_2O	Lung mechanics

Here *m* indicates the respiratory cycle.

TABLE 10.2 Input and Output Variables of the Control of Breathing Model

depends on the partial pressures of oxygen and carbon dioxide in arterial blood, ignoring the effect of these blood gases in other locations, and the effect of other blood components, such as pH, and the lung stretch receptors (Guyton, 2006).

10.2 Conceptual Models

The basic principle underlying the models presented in this chapter is linearization of the empirical relationship between the controlled variable and the control effector in the operating point (Fig. 10.1). This results in a "gain with offset" model (Table 4.11). A piecewise linear function (PLF) (also see Table 4.11) and limits are included to expand the range of validity of the model. The PLF represents receptor activation and saturation as a function of the controlled variable. Overall physiological limits are imposed on the control effector. Figure 10.3 gives an example of these relationships in baroreflex control of heart rate.

In this model, both saturations of the *hr*(*map*) curve [Fig. 10.3(c)] result from the saturation of the receptor [Fig. 10.3(a)], not from the limits on the control effector. The baroreceptor activation function is common to all four cardiovascular control effectors, but the gains and control effector saturations are distinct.

Control of breathing is based on two receptor activation functions, one for each of the controlled variables p_aO_2, and p_aCO_2. Oxygen has a stimulating effect on spontaneous breathing for $p_aO_2 <$ 100 mm Hg, but no effect for higher values. High carbon dioxide, $p_aCO_2 >$ 40 mm Hg, has a stimulating effect; low carbon dioxide has a depressing effect. The parameters governing these effects will be presented in the next section. We assume that the main effects of the controlled variables on spontaneous breathing are on minute ventilation, the product of tidal volume and respiratory rate, leaving the ratio between the two intact. These effects are governed by two gains and are considered additional. Physiological limits are applied to tidal volume and respiratory rate. For a particular set of lung mechanics parameters, and in the approximation of the lung mechanics model presented in the previous chapter, tidal volume is linearly related to the output variable muscle pressure amplitude.

Simplifying assumptions included in these conceptual models are that for the intended purpose, the relationships between controlled variables and control effectors can indeed be represented by static, piecewise linear relationships. The additive nature of the effect of p_aO_2 and p_aCO_2 on minute ventilation is another simplifying assumption. More elaborate conceptual and mathematical models of control of circulation, control of respiration, and their interactions can be found in, for example, the work by Ursino (1998), and Khoo (1999).

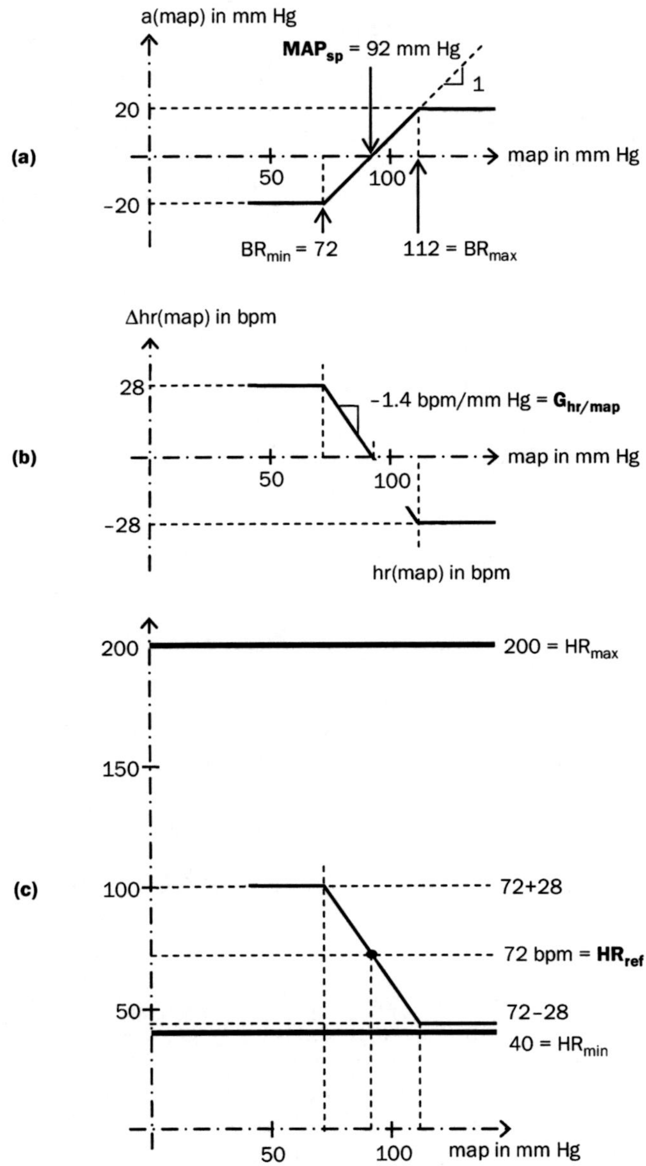

FIGURE 10.3 Relationships in baroreflex control of heart rate. (a) Baroreceptor activation function *a*(*map*). (b) Baroreflex-mediated heart rate change $\Delta hr(map) = G_{hr/map} * a(map)$. (c) MAP-dependent heart rate $hr(map) = HR_{REF} + \Delta hr(map)$ and saturations. Numerical values as in Tables 10.3, 10.4, and 10.5.

10.3 Mathematical Models

For notational compactness, we drop the (discrete) time notation in the static relationships between controlled variables and control effectors. The distinction between variables and parameters is maintained via lower and upper case symbols. The receptor activation function of control of circulation [Fig. 10.3(a)] can be described as follows:

$$a_{map}(map) = \begin{cases} BR_{min} - MAP_{sp} & \text{if} \quad map \leq BR_{min} \\ BR_{max} - MAP_{sp} & \text{if} \quad map \geq BR_{max} \\ map - MAP_{sp} & \text{otherwise} \end{cases} \tag{10.1}$$

with the mean arterial pressure setpoint MAP_{sp}, the threshold below which decrease in *map* has no further effect on the baroreceptor BR_{min}, and the saturation point of baroreceptor activation BR_{max}. This function is multiplied by the gain from *map* to heart rate $G_{hr/map}$ [also see Fig. 10.3(b)].

$$\Delta hr = G_{hr/map} a_{map}(map) \tag{10.2}$$

Addition of the heart rate reference HR_{ref} and control effector saturation [Fig. 10.3(c)] is accomplished as follows:

$$s_{hr}\left(HR_{ref} + \Delta hr\right) = \begin{cases} HR_{min} & \text{if} \quad HR_{ref} + \Delta hr \leq HR_{min} \\ HR_{max} & \text{if} \quad HR_{ref} + \Delta hr \geq HR_{max} \\ HR_{ref} + \Delta hr & \text{otherwise} \end{cases} \tag{10.3}$$

Other control effectors are computed in a similar way, based on the same receptor activation function $a_{map}(map)$, but with different references MC_{ref}, SAR_{ref}, and VUV_{ref}; gains $G_{mc/map}$, $G_{sar/map}$, and $G_{vuv/map}$; and control effector saturation functions $s_{mc}()$, $s_{sar}()$, and $s_{vuv}()$. The parameters of these saturation functions are MC_{min}, MC_{max}, SAR_{min}, SAR_{max}, VUV_{min}, and VUV_{max}, respectively. This relatively simple empirical model has many parameters, which can be grouped as follows. Table 10.3 lists the parameters of the operating point. The first four are in fact parameters of the hemodynamic model, and MAP_{sp} is the value of the hemodynamic output *map*(*n*) for an unperturbed system in open-loop. "Open-loop" is interpreted here in the sense of the defined control model: There is no feedback of deviations in the controlled variable from its setpoint to the control

Name	Symbol	Value	Unit	Table 8.3
Heart rate reference	HR_{ref}	72	min^{-1}	$60/HP$
Myocardial contractility reference	MC_{ref}	4	mm Hg ml^{-1}	E_{lvmax}
Systemic arteriolar resistance reference	SAR_{ref}	0.8	mm Hg ml^{-1} s	R_{sart}
Venous unstressed volume reference	VUV_{ref}	2190	ml	$UV_{etv} + UV_{itv}$
Mean arterial pressure setpoint	MAP_{sp}	92	mm Hg	

Table 10.3 Operating Point of the Unperturbed Hemodynamic Model in Open-Loop

effectors. Physiologically speaking, the autonomic nervous system may be active, even in the operating point. The table also lists corresponding parameters and numerical values included in Table 8.3.

Gains from *map* to the control effectors of the Wesseling and Settels model, as presented by Ten Voorde (1992), and used in Goodwin et al. (2004), are listed in Table 10.4.

As all the change in total peripheral resistance are assumed to be concentrated in the systemic arterioles, the gain $G_{sar/map}$ is equal to the gain $G_{TPR/map}$. With the control as defined by Eqs. (10.1) through (10.3), if the control loop is closed on an unperturbed system, the controlled variable should remain in its setpoint, and the control effectors should keep their reference values.[2] Parameters of the receptor activation function and control effector saturations are listed in Table 10.5.

The baroreceptor thresholds are set so that, in combination with the gain $G_{hr/map}$, they result in a reasonable baroreflex mediated change in heart rate. The range of mean arterial pressures with an influence on the firing frequency of nerves coming from the baroreceptors is larger, but nonlinear (Guyton, 2006). The limits imposed on the control effectors are empirically set general physiological limits for a healthy adult male, not only reflecting the influence of the baroreflex.

Control of breathing is modeled analogously, via oxygen and carbon dioxide receptor activation functions causing changes in (exhaled) minute ventilation via dedicated gains.

$$\Delta ve = G_{ve/paO2} a_{paO2}(p_a O_2) + G_{ve/paCO2} a_{paCO2}(p_a CO_2) \tag{10.4}$$

[2] High gains and system time constants could result in sustained oscillations.

Name	Symbol	Value	Unit
Map-heart rate gain	$G_{hr/map}$	−1.4	min^{-1} mm Hg^{-1}
Map-myocardial contractility gain	$G_{mc/map}$	−0.016	ml^{-1}
Map-systemic arteriolar resistance gain	$G_{sar/map}$	−0.027	s ml^{-1}
Map-venous unstressed volume gain	$G_{vuv/map}$	27	ml mm Hg^{-1}

TABLE 10.4 Gains of the Control of Circulation Model

Name	Symbol	Value	Unit
Baroreceptor lower threshold	BR_{min}	72	mm Hg
Baroreceptor saturation	BR_{max}	112	mm Hg
Minimum heart rate	HR_{min}	40	min^{-1}
Maximum heart rate	HR_{max}	200	min^{-1}
Minimum myocardial contractility	MC_{min}	0	mm Hg ml^{-1}
Maximum myocardial contractility	MC_{max}	12	mm Hg ml^{-1}
Minimum systemic arteriolar resistance	$SART_{min}$	0.2	mm Hg ml^{-1} s
Maximum systemic arteriolar resistance	$SART_{max}$	2.4	mm Hg ml^{-1} s
Minimum venous unstressed volume	VUV_{min}	1000	ml
Maximum venous unstressed volume	VUV_{max}	4000	ml

TABLE 10.5 Activation and Saturation Function Parameters of the Control of Circulation Model

Assuming a constant ratio between tidal volume and respiratory rate of $500/12 \approx 41.7$ ml min, tidal volume and respiratory rate can be computed from minute ventilation.

$$vt = \sqrt{\frac{500}{12}\left(VE_{ref} + \Delta ve\right)} \tag{10.5}$$

$$rr = \sqrt{\frac{12}{500}\left(VE_{ref} + \Delta ve\right)} \tag{10.6}$$

Physiological limits are applied to these two variables. The output variable muscle pressure amplitude is computed from tidal volume via the inverse total lung and chest wall compliance (Table 9.3).

$$mpa = \frac{1}{110} vt \tag{10.7}$$

This computation ignores the dynamic nature of the breathing cycle. Table 10.6 lists the parameters of the operating point of the respiratory system. Via Eq. (10.7) the tidal volume reference corresponds to the muscle pressure amplitude in Table 9.3. The rate reference corresponds to the respiratory period in the same table. The setpoints should correspond to the values of corresponding outputs of the unperturbed lung mechanics and uptake and distribution models in open-loop.[3]

Table 10.7 lists the gains of the control of breathing model. These gains are set empirically with respect to the minute ventilation in the operating point so that a decrease of 5 mm Hg in p_aCO_2 leads to apnea. The response to oxygen is half as strong as the carbon dioxide response, and in the opposite direction (a drop in oxygen leads to hyperventilation).

Table 10.8 lists the activation and saturation function parameters of the control of breathing model.

Name	Symbol	Value	Unit
Tidal volume reference	VT_{ref}	500	ml
Respiratory rate reference	RR_{ref}	12	min^{-1}
Partial pressure of oxygen in arterial blood.	P_aO_{2sp}	100	mm Hg
Partial pressure of carbon dioxide in arterial blood.	P_aCO_{2sp}	40	mm Hg

TABLE 10.6 Operating Point of the Unperturbed Respiratory Models in Open-Loop

Name	Symbol	Value	Unit
Oxygen-minute ventilation gain	$G_{ve/paO2}$	−600	ml min^{-1} mm Hg^{-1}
Carbon dioxide-minute ventilation gain	$G_{ve/paCO2}$	1200	ml min^{-1} mm Hg^{-1}

TABLE 10.7 Gains of the Control of Breathing Model

3 Verified for similar models in the METIman™ software.

Name	Symbol	Value	Unit
Oxygen receptor lower threshold	P_aO_{2min}	20	mm Hg
Oxygen receptor saturation	P_aO_{2max}	100	mm Hg
Carbon dioxide receptor lower threshold	P_aCO_{2min}	20	mm Hg
Carbon dioxide receptor saturation	P_aCO_{2max}	70	mm Hg
Minimum tidal volume	VT_{min}	0	ml
Maximum tidal volume	VT_{max}	4600	min^{-1}
Minimum respiratory rate	RR_{min}	0	min^{-1}
Maximum respiratory rate	RR_{max}	40	min^{-1}

Table 10.8 Activation and Saturation Function Parameters of the Control of Breathing Model

Unlike the baroreceptor activation function, the modeled p_aO_2 and p_aCO_2 receptor activation functions are not symmetrical around the setpoint. The lower threshold on the oxygen receptor is set equal to the setpoint. This eliminates the oxygen response for $p_aO_2 > 100$ mm Hg; high oxygen does not cause hypoventilation. The limits imposed on the control effectors are again empirically set general physiological limits for a healthy adult male, not only reflecting the influence of control of breathing.

References and Further Reading

Chiari L, Avanzolini G, Ursino M. A comprehensive simulator of the human respiratory system: validation with experimental and simulated data. *Ann Biomed Eng.* Nov-Dec 1997; 25(6):985–99.

Goodwin JA, van Meurs WL, Sá Couto CD, Beneken JEW, Graves SA. A model for educational simulation of infant cardiovascular physiology. *Anest Analg.* Dec 2004; 99(6):1655–64.

Guyton AC, Hall JE. *Textbook of medical physiology.* 11th ed. Philadelphia: W.B. Saunders; 2006.

Khoo MCK. *Physiological control systems, analysis simulation, and estimation.*, New York: Wiley-IEEE Press; 1999.

Ursino M. Interaction between carotid baroregulation and the pulsating heart: a mathematical model. *Am J Physiol Heart Circ Physiol.* 1998; 275: H1733–H1747.

Ten Voorde B. Modeling the baroreflex: a system analysis approach [thesis]. Amsterdam, The Netherlands: Vakgroep Medische Fysica; 1992.

Van de Vegte J. *Feedback control systems.* 3rd ed. Englewood Cliffs, NJ: Prentice Hall; 1994.

Wesseling KH, Settels JJ. Baromodulation explains short-term blood pressure variability. In: Orlebeke JF, Mulder G, van Doornen LJP, eds. *Psychophysiology of cardiovascular control.* New York: Plenum Press; 1985:69–97.

Zijlmans M, Sá-Couto CD, van Meurs WL, Goodwin JA, Andriessen P. Corrected and improved model for educational simulation of neonatal cardiovascular pathophysiology [technical report]. *Simul Healthc.* Spring 2009; 4(1):49–53.

PART III

Advanced Topics

CHAPTER 11

Sensitivity Analysis

In this chapter we describe a method for establishing relative sensitivity of model outputs to changes in model parameters, and apply it to a selection of model parameters and outputs of the hemodynamic model introduced in Chap. 8. Physiological consistency of the results is discussed.

A sensitivity analysis basically answers the questions: "What is affected by what, and by how much?" We can consider the sensitivity of output variables to input variables, parameters, and initial conditions. Sensitivity analysis has been applied to physiological models for evaluation of the relevance of a particular parameter to the model response, for reduction of model complexity, and to assist in parameter estimation. In Sec. 11.1 we introduce a method for establishing relative sensitivity of model outputs to changes in model parameters. In the subsequent section we apply this technique to the uncontrolled hemodynamic model introduced in Chap. 8, with the specific purpose to assist users of a model-driven training simulator in programming new patients.

11.1 Method

Consider a mathematical model with M parameters P_i, $i = 1, 2, \dots, M$, and N output variables v_j, $j = 1, 2, \dots, N$. Let $P_{i,0}$ be the baseline value for parameter P_i, and $v_{j,0}$ the baseline value for variable v_j. A variation ΔP_i in P_i potentially causes variations Δv_j in all variables v_j. For a dynamic model, steady-state conditions need to be considered. The relative sensitivity of the output variable v_j to the parameter P_i is defined as:

$$S_{ij} = \frac{\Delta v_j / v_{j,0}}{\Delta P_i / P_{i,0}} \tag{11.1}$$

The values S_{ij} constitute an $M \times N$ matrix, called the relative sensitivity matrix. The elements of this matrix are easily interpreted as the relative change in an output variable in response to a relative change in a parameter. For example, the relative sensitivity of systolic blood pressure (SBP) to heart rate (HR) in the uncontrolled hemodynamic model presented in Chap. 8 is 0.26. This means that a 10% increase in HR will lead to an approximately 2.6% increase in SBP. The normalized, dimensionless elements in the sensitivity matrix can be compared among each other, even if the output variables have different units. For example, the relative sensitivity of cardiac output (CO) to HR in the same model is 0.41. In this model, CO is more sensitive to HR than SBP.

For simple models, the sensitivity matrix may have an analytical solution, obtained via differentiation. For nonlinear, dynamic models, it is often determined via simulation experiments. The results presented in the next section are obtained with $\Delta P_i / P_{i,0} = 0.01$; that is, simulations are carried out to evaluate the effect of a positive 1% change in each parameter, one at a time. The size of this change has to be carefully balanced: Too small a parameter change may result in truncation errors in the subtractions involved in computing the changes in the variables, whereas too big a parameter changes may result in nonlinear effects. For a nonlinear model, the sensitivity matrix also depends on the particular operating point. Additional information can be obtained by modifying the magnitude of the parameter change, or its sign, and by combining changes in parameters.

When interpreting relative sensitivities, it has to be noted that typical changes in parameters may differ greatly. For example, the sensitivities of flow rates and blood pressures to changes in systemic vascular resistance (SVR) are much higher than the sensitivities to changes in cardiac valve resistances; doubling SVR results in considerable physiological changes, whereas doubling of the resistance of a cardiac valve has no observable effects at all. Yet, in valve disease, the increase in resistance may be as much as hundredfold, resulting in a significant effect. Another limit associated with relative sensitivity is that it amplifies absolute changes in variables with a low baseline value. For example, the relative sensitivity of central venous pressure to parameter changes is amplified by an average baseline value close to 0 mm Hg.

11.2 Application to a Hemodynamic Model

The sensitivity analysis introduced in the previous section is applied to selected parameters and output variables of the hemodynamic model introduced in Chap. 8. This analysis involves the unperturbed model in its open-loop operating point (also see Chap. 10). We will refer to this model as the Beneken model. This model was also the basis of the cardiovascular model used in the Human Patient Simulator (HPS®), developed at the University of Florida, and commercialized by Medical Education Technologies, Inc. In the context of the development of a simulator-based hemodynamic monitoring class (Öhrn, 1995), HPS® developers introduced six normalized hemodynamic "factors" grouping selected model parameters that can be used to adapt an existing baseline patient to reflect a specific physiology or pathology. The relationship between these factors and the corresponding parameters in the Beneken model is reflected in Tables 11.1a and b. These tables present the quantitative results of a sensitivity analysis, as well as a condensed view, indicating only the sign and the number of significant digits in

Beneken model parameter	SBP	DBP	SPAP	DPAP	CO	HPS® factor
HR	0.26	0.54	−0.15	0.46	0.41	Heart rate
E_{lvmax}	0.09	0.06	−0.02	−0.16	0.07	Left ventricular contractility
E_{rvmax}	0.15	0.15	0.19	0.23	0.16	Right ventricular contractility
R_{sart}	0.41	0.67	−0.10	−0.05	−0.14	SVR
$0.8\ R_{pt}$	−0.07	−0.07	0.31	0.73	−0.07	PVR
$UV_{esv}+UV_{isv}$	−1.31	−1.33	−1.53	−1.99	−1.27	Venous capacity

SBP, systolic blood pressure; DBP, diastolic blood pressure; SPAP, systolic pulmonary artery pressure; DPAP, diastolic pulmonary artery pressure; CO, cardiac output.

Table 11.1a Sensitivity Analysis, Quantitative Results

Beneken model parameter	SBP	DBP	SPAP	DPAP	CO	HPS® factor
HR	++	++	−−	++	++	Heart rate
E_{lvmax}	+	+	−	−−	+	Left ventricular contractility
E_{rvmax}	++	++	++	++	++	Right ventricular contractility
R_{sart}	++	++	−−	−	−−	SVR
$0.8\ R_{pt}$	−	−	++	++	−	PVR
$UV_{esv} + UV_{isv}$	−−−	−−−	−−−	−−−	−−−	Venous capacity

Table 11.1b Sensitivity Analysis, Condensed View

the result, respectively. The results of this sensitivity analysis apply most directly to the baseline patient called "Stan Vintage" (van Meurs, 2006).

These results are consistent with a more extensive sensitivity analysis of the same model presented by Sá Couto and van Meurs (2009). Most of the sensitivities are fairly intuitive; increases in both *HR* and SVR (R_{sart}) result in increases in SBP and DBP, but an increase in *HR* results in an increase in CO, whereas an increase in SVR results in a decrease in CO. The strong effect of venous capacity on the whole circulation and the effect of right ventricular contractility on both pulmonary and systemic pressures are noteworthy. The negative sensitivity of SPAP to *HR* may be due to a redistribution of blood volume between the pulmonary and systemic parts of the circulation, and is consistent with a reduction in CVP following the increase in *HR*. The results of the sensitivity analysis can also be interpreted in the opposite direction: What are the parameters affecting a particular variable? For example, CO is strongly depressed by an increase in venous capacity (venodilation) and strongly stimulated by an increase in *HR*. In this model and this operating point, the effect of left ventricular contractility on CO is rather limited.

References and Further Reading

Öhrn MAK, van Meurs WL, Good ML. Laboratory classes: replacing animals with a patient simulator [abstract]. *Anesthesiology*. 1995; 83(3A):A1028.

Sá Couto CD, van Meurs WL. Sensitivity analysis of a hemodynamic model to facilitate programming simulation patients. in: Sá Couto CD. Mathematical models for educational simulation of cardiovascular pathophysiology [thesis]. Porto, Portugal: University of Porto College of Engineering; 2009.

Van Meurs WL, Neto P, Azevedo H, Sá Couto CD. "Stan Vintage": A baseline patient for the Human Patient Simulator with hemodynamic parameters from the scientific literature [abstract]. *Simul Healthc*. 2006; 1(3):183.

CHAPTER 12

Design of Model-Driven Acute Care Training Simulators

A simulation-based training program is designed to meet carefully specified training needs. In the case of full-mission simulations for training high-performance tasks we strive for a high level of functional fidelity, which pertains to the similarity of trainee behavior in the simulated and real environments. Physical fidelity is the extent to which the appearance and behavior of a simulator match the appearance and behavior of the simulated system. Sufficient physical fidelity is achieved if a designed simulator allows meeting the specified training needs at the desired level of functional fidelity. Model requirements are set, and operational validity established, based on the required simulator behavior.

A physiologic model is rarely designed and evaluated in isolation; its requirements are embedded in more general project goals, and evaluation of conceptual and operational validity (Chap. 6) refers back to these requirements. The purpose of this chapter is to illustrate how model requirements can be derived from more general process and device requirements. We will use simulation-based training in anesthesia as an example.

Farmer et al. (1999) observe that the most critical skills acquired with the aid of military training simulators are skills required for the performance of so-called *high-performance tasks*. These tasks are defined as complex, time-critical tasks where the operator is in the primary control loop. Farmer et al. further define a *full-mission simulator* as a top-level training device that provides all the required cues and facilities necessary for training complete missions (of a specific weapon system). Many tasks in acute care medicine (anesthesia, intensive care, emergency medicine, and labor and delivery) are high-performance tasks, and, starting with the specialty of anesthesia, several full-mission simulations and simulators were developed (Denson, 1969; Gaba, 1988; Good, 1989; Chopra, 1994; Rettedal, 1996).

Training in a simulated acute care environment has a number of advantages over training in the clinical environment. First, the real patient, or an animal model, is replaced by a simulator. Errors can be made and learned from, without any other consequences; rare critical events can be trained on demand; and the pace of training can be adapted to the progress of the trainees. However, realism in the simulated environment is necessarily limited, and often comes at a cost. Some discrepancies between the simulated and real environments may be acceptable, and can be overcome by soliciting the imagination of trainees, who are capable of remarkable leaps of faith, provided they buy into the simulation. Other discrepancies could impede meeting the educational goals. The design of simulation-based training programs and training simulators faces the challenge of anticipating what needs to be included and what can be left out, while keeping the design pedagogically coherent, and technically and economically feasible.

Two concepts can provide guidance in simulation and simulator design. *Functional fidelity* pertains to the similarity of trainee behavior in the simulated and real environments. *Physical fidelity* is the extent to which the appearance and behavior of a simulator match the appearance and behavior of the simulated system. Farmer et al. present a detailed procedure for the design and evaluation of training programs and simulators. In the next two sections we briefly present this procedure and illustrate how it can be applied in health care simulation, backtracking certain aspects of the design of full-mission anesthesia simulations and simulators.[1] Specification of model requirements fits in this procedure, and is the topic of the final section.

12.1 Training Program Design

The design of a training program is based on the results of a training needs analysis (TNA).[2] Table 12.1 lists the main steps in TNA as specified by Farmer et al. and provides a few examples of what is subject to analysis for the target learners: anesthesia residents.

A training program can now be designed to meet the identified training needs. Table 12.2 lists the main steps in training program

[1] In many practical situations, simulation program design will follow, rather than precede, simulator design, and be carried out by different (teams of) designers using available training media. However, the principle of letting educational considerations drive simulator evaluation and procurement also applies in these situations.

[2] In this chapter we will not consider the more global objectives of the health care system and overall curriculum.

	Examples
Mission analysis	Airway management and mechanical ventilation for general anesthesia
Task analysis	Machine check, preoxygenation, medication, laryngoscopy and tracheal intubation, ventilator operation, weaning and extubation, troubleshooting: difficult airway, breathing circuit problems
Trainee analysis	Performance on an intubation part-task trainer
Training analysis	Formulation of skill tests

Table 12.1 Main Steps in Training Needs Analysis

	Examples
Specification of training activities	Briefing, normal intubation and mechanical ventilation, airway complication, breathing circuit problem, debriefing
Scenario design	Cannot intubate/cannot ventilate, breathing circuit disconnect
Design of instruction	Prompting by a facilitator?

Table 12.2 Main Steps in Training Program Design

design (TPD). Background information for TNA and TPD can be found in Bready (2007), or in more traditional anesthesia textbooks and guidelines.

In the case of full-mission simulations for training high-performance tasks, we strive for a high level of functional fidelity, under the assumption that this facilitates transfer of skills acquired in the simulated environment to the real environment. Any TPD decision that brings the (anticipated) behavior of the trainee in the simulated environment closer to behavior in the real environment increases functional fidelity. Other factors that contribute to functional fidelity include competent instructors using appropriate training media. An important aspect of training media specification (TMS), simulator design, is presented in the next section.

Functional fidelity is a useful simulation design concept, but has limited application in training evaluation (TE), mainly because it does not reflect endpoints, such as increased task performance or improved patient outcome. Preimplementation evaluation of a proposed training program can be achieved by submitting the content to domain experts. After implementation of a simulation-based training program, TE can focus on endpoints, for example, by assessing the performance of trainees on standardized skill tests. Sometimes improved health care in terms of patient outcomes can be demonstrated, see, for example, Draycott et al. (2006). More subjective forms of TE include check lists filled out by instructors and self-evaluation and satisfaction scores by trainees.

12.2 Simulator Design

Dieckmann et al. (2007) observed that it is not uniformly the case that more physical fidelity means better attainment of educational goals. However, the level of physical fidelity needs to be sufficient to meet the specified training needs at the desired level of functional fidelity in a majority of trainees. As observed in the previous section, in the case of full-mission simulations for training high-performance tasks, the desired level of functional fidelity is high. Simulators supporting these objectives should therefore have a sufficiently high level of physical fidelity.

Before discussing simulator design, we briefly introduce the functional units of acute-care simulators. All but one of the full-mission acute-care simulators listed in the introductory section to this chapter, and many part-task trainers, are script-controlled, model-driven devices (van Meurs, 1997). Figure 12.1 shows the main functional units in such simulators.

The trainees interact with the simulator via a number of interfaces for clinical signs, monitored signals, and therapeutic interventions. The patient mannequin and medical devices can be real, or emulated on a computer screen. The instructor, sometimes assisted by an operator, has run-time, and possibly off-line control over the simulation. The instructor can make the simulated patient evolve or react to therapeutic interventions, but in that is often assisted by a combination of time-and-event-based scenario

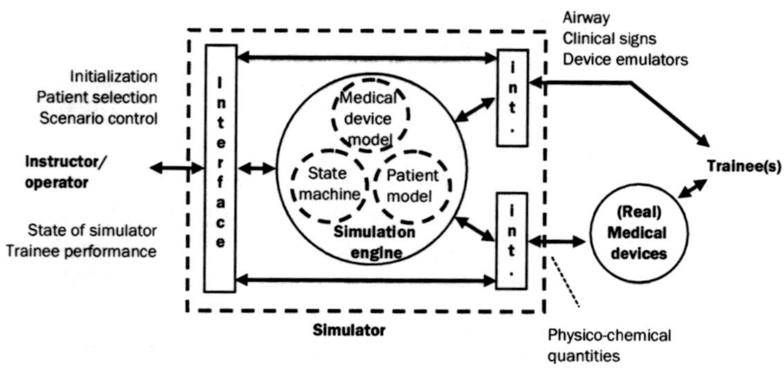

FIGURE 12.1 Functional units and run-time information flow in a script-controlled model-driven simulator. (*Adapted from van Meurs, 2003.*)

			Examples
Interfaces	Trainee interfaces	Mannequin	Airway, breath sounds
		Monitors	Capnograph
		Therapeutic devices	Mechanical ventilator
	Instructor interfaces	Run-time	Patient and scenario selection
		Off-line	Log of trainee activities
Simulation engine	Models	Patient	Respiratory system
		Devices	Gas mixing in ventilator
	Scripts		Triggering airway occluder

Table 12.3 Functional Units in an Acute Care Training Simulator

scripts, running on a finite state machine, and continuous-time models. Table 12.3 further divides these functional units and provides examples of what needs to be incorporated to create a simulator implementing the training program outlined in the previous section.

Appearance and behavior of the trainee interfaces are subject to design and directly influence physical fidelity. For example, to optimize physical and functional fidelity we decide that a full-body mannequin and a real mechanical ventilator and real monitoring equipment will be used. This implies that the simulator needs to interface with these devices. It also implies that a device model for gas mixing in the ventilator is not required. Another example is that breath sounds need to appear in appropriate locations, and need to be bilateral so that endobronchial intubation[3] can be diagnosed.

Appearance and behavior of the instructor-operator interface are also subject to design, but only indirectly influence physical fidelity. However, careful design can enhance ease of operation, and inclusion of learning management tools can increase the impact of a simulation. Examples of off-line instructor interface tools include patient and scenario editors. Both run-time and off-line instructor interfaces may interact with physiologic models, and specific model requirements may result. The concept of hemodynamic factors, described in Sec. 11.2, constitutes an example.

The simulation engine is the main determinant of simulator behavior, and an important factor in physical fidelity. For example, to

[3] Positioning of the tip of an endotracheal tube beyond the split of the trachea in one of the mainstem bronchi leads to ventilation of a single lung.

simulate airway management and mechanical ventilation, alveolar oxygen and carbon dioxide should depend in a realistic fashion on the timing and precision of tracheal intubation, and on tidal volume and respiratory rate set on a mechanical ventilator. Some of the dynamic aspects of patient evolution and responses to therapy can be taken into account by a script. For example, an airway occluder, actuated by the instructor or by a scenario script, is required to implement the cannot ventilate/cannot intubate scenario. However, a scripted simulation cannot be designed to anticipate all possible management options, nor generate all desired output signals, especially in the case of multiple, continuous independent and dependent variables. Therefore, this requirement leads to the adoption of an at least partially model-driven simulation engine. Model requirements will be discussed further in the next section.

In many cases, the extent to which the appearance and behavior of a simulator match the appearance and behavior of the simulated system is a subjective notion, and can only be evaluated qualitatively. Another design challenge results from the interdependence of interface and simulation engine design. For example, for realistic simulation of endotracheal intubation, the anatomy of the airway needs to be correct, breath sounds need to appear in the right locations and be triggered appropriately, and physiologic responses need to be suitable.

Other considerations in simulator design include economic and technological feasibility, and safety, cost, and ease of operation and maintenance. The effect of any remaining discrepancies between the appearance and behavior of simulated and real patients and equipment can be mitigated by addressing them in a simulation briefing. Elaborating on our comment in the previous section, functional and physical fidelity are useful concepts in the simulation and simulator design, but do not reflect endpoints in the effect of training. Training evaluation (TE) following simulator development and training program implementation remains the primary tool to demonstrate adequacy (or insufficiency) of simulations and simulators.

12.3 Model Requirements

Many acute-care simulation projects have in common the requirement to provide realistic and consistent evolution of clinical signs and monitored signals, and of their response to therapeutic interventions (Zijlmans, 2009). A simulation engine can generate these variables automatically and in real time. For complex systems this simulation engine needs to include models. Realistic reactions require appropriate interactions between physiological subsystems. For example, in the case of mechanical ventilation with higher airway pressures, the circulation should become progressively compromised. This requirement leads to a gray-box modeling approach, where the resulting models reflect structure and function, facilitating coupling of models of physiological subsystems. In educational simulation it is often required to adapt the physiology of the patient to simulate pathologies or critical incidents. This requirement further reinforces the preference for a gray-box modeling approach, so that models can be manipulated in a logical fashion to achieve the desired goals.

Following the format of Table 2.1, qualitative requirements for a gray-box model, part of a simulator for training airway management and mechanical ventilation for general anesthesia, are listed in Table 12.4.

The specific examples listed in Table 12.4 can all be traced back to simulator or simulation scenario requirements presented in the previous sections. Although the main training objectives are airway management and mechanical ventilation, simulation of induction of anesthesia also requires a reasonably realistic response of both the cardiovascular and respiratory systems to anesthetic agents. An

	Examples
Organ or physiological subsystem	Lung mechanics, uptake and distribution of respiratory gases, effect of anesthetic agents
Population	ASA PS 1, 2,* patient with a difficult airway.
Physiological states	Normal
Pathologies	None
Clinical signs and monitored variables	Breath sounds, capnogram, pulse oximetry
Critical incidents	Cannot intubate/cannot ventilate, breathing circuit disconnect
Interventions	Preoxygenation, medication, ventilator settings
Overall block diagram	See main text
Internal structures and functions	To be determined

*American Society of Anesthesiologists (ASA) Physical Status (PS) reflecting normal healthy patients and patients with mild systemic disease.

Table 12.4 Qualitative Model Requirements

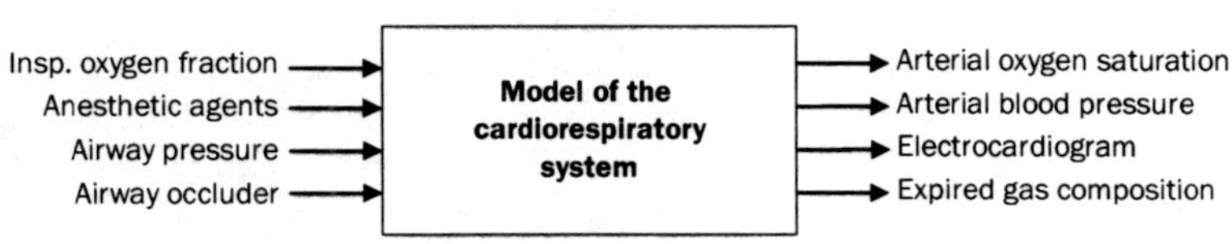

Figure 12.2 Overall block diagram of the cardiorespiratory model.

overall block diagram of the cardiorespiratory model is given in Fig. 12.2.

The output variables include signals shown on standard monitors for induction of anesthesia. The model requirements can be met using a mathematical model and model development can follow the process outlined in this book. Alternatively, models can be implemented using both mathematical and physical components. Such hybrid models are difficult to design, but have the advantage that the interaction with, for example, a real ventilator and gas analyzer takes place via physical quantities, rather than via computer generated variables. For example, a hybrid lung mechanics model, using real gas flows and an anatomically correct airway (Sajan, 1993), can easily detect endobronchial intubation or insufficient sealing of the airway, thus increasing physical fidelity. Note that these phenomena are difficult to observe otherwise.

In the previous sections we observed that in the case of full-mission simulations for training high-performance tasks we strive for a high level of functional fidelity. The level of physical fidelity of a simulator needs to be sufficient to meet the specified training needs at the desired level of functional fidelity. The simulation engine is the main determinant of simulator behavior, and an important factor in physical fidelity. We recall from Chap. 6 that a model is considered operationally valid for a set of experimental conditions if its accuracy is acceptable for the intended purpose. This is a slightly more formal way of looking at one of the components of physical fidelity: The extent to which the behavior of a simulator matches the behavior of the simulated system. Target data for a representative set of clinical situations, likely to be encountered in a simulation-based training exercise, form the basis of model validation. What is acceptable in view of the educational objectives may remain a matter of informed opinion. We also recall from Chap. 6 that target data for the rare and dangerous situations which are most interesting in simulations for training high-performance tasks, obtained on humans in carefully controlled conditions, are often not available. Conceptual validity refers to the acceptability of the incorporated theories, assumptions, and data. High-quality gray-box models used in training simulators facilitate coupling of models of physiological subsystems and manipulation of models to achieve the desired educational goals.

The extensive example treated in this chapter demonstrates that a physiologic model is rarely designed and evaluated in isolation; it also illustrates how model requirements can be derived from more general process and device requirements. It should be noted that existing full-mission healthcare simulators were designed in much less structured, more intuitive, ways. The presented top-down design process has too many assumptions and interplaying factors to evolve into a true "science of acute care simulator design." However, it may provide guidance to the design of new simulations, simulators, and models, and to the adaptation of existing ones, and its application should increase the traceability of design decisions at several levels to educational objectives.

References and Further Reading

Bready LL, Dillman D, Noorily SH, eds. *Clinical decision making in anesthesiology*, 4th ed. Philadelphia: Mosby Elsevier; 2007.

Chopra V, et al. The Leiden anaesthesia simulator. *Br J Anaesth* 1994; 73:287–92.

Denson JS, Abrahamson S. A computer-controlled patient simulator. *JAMA*. 1969; 208:504–8.

Dieckmann P, Gaba D, Rall M. Deepening the theoretical foundations of patient simulation as social practice. *Simul Healthc*. Fall 2007; 2(3):183–93.

Draycott T, Sibanda T, Owen L, Akande V, Winter C, Reading S, Whitelaw A. Does training in obstetric emergencies improve neonatal outcome? *BJOG*. Feb 2006; 113(2):177–82.

Farmer E, van Rooij J, Riemersma J, Jorna P, Moraal J. *Handbook of simulator-based training*. Aldershot: Ashgate; 1999.

Gaba DM, DeAnda A. A comprehensive anesthesia simulation environment: re-creating the operating room for research and training. *Anesthesiology*. 1988; 69:387–94.

Good ML, Gravenstein JS. Anesthesia simulators and training devices. *Int Anesthesiol Clin* 1989; 27(3):161–8.

Maran NJ, Glavin RJ. Low- to high-fidelity simulation – a continuum of medical education? *Med Educ*. Nov 2003; 37 Suppl 1:22–8.

Rettedal A, Stale F, Ragna K, Petter L. PatSim – simulator for practicing anaesthesia and intensive care, development and observation. *Int J Clin Monit Comput*. 1996; 13:147–52.

Sajan I, van Meurs WL, Lampotang S, Good ML, Principe JC. Computer controlled mechanical lung model for an anesthesia simulator, abstracted. *Int J Clin Monit Comput*. 1993; 10:194–5.

Van Meurs WL, Good ML, Lampotang S. Functional anatomy of full-scale patient simulators. *J Clin Monit*. 1997; 13(5): 317–24.

Van Meurs WL, Sá Couto PM, Sá Couto CD, Bernardes JF. Ayres-de-Campos D. Development of foetal and neonatal simulators at the University of Porto. *Med Educ*. 2003; 37 Suppl 1:29–33.

Zijlmans M, Sá-Couto CD, van Meurs WL, Goodwin JA, Andriessen P. Corrected and improved model for educational simulation of neonatal cardiovascular pathophysiology [technical report]. *Simul Healthc*. Spring 2009; 4(1):49–53.

Index

Page numbers followed by t or f indicate tables or figures, respectively

H

I

K

L

M

CPSIA information can be obtained at www.ICGtesting.com
Printed in the USA
LVOW09*1135101215

466259LV00010B/43/P